Aldo Leopold

The Man and His Legacy

Thomas Tanner, Editor
Foreword by Curt Meine

SOIL
AND WATER
CONSERVATION
SOCIETY

Soil and Water Conservation Society
945 SW Ankeny Road
Ankeny, Iowa 50023

Soil and Water Conservation Society
945 SW Ankeny Road, Ankeny, IA 50023
www.swcs.org

This edition includes the Forewords written by Stewart L. Udall in 1987 and by Paul W. Johnson in 1995.

Images on the cover are courtesy of the Aldo Leopold Foundation.

ISBN 978-0-9856923-0-8

The Library of Congress has cataloged the previous edition as follows:

Library of Congress Catalog Card Number 95-22547

Aldo Leopold: The man and his legacy / Thomas Tanner, editor;
foreword by Paul W. Johnson.
p.   cm.
"50th anniversary edition."
Includes bibliographical references.
ISBN 0-935734-38-4
1. Leopold, Aldo, 1886-1948.
2. Naturalists-Wisconsin-Biography.
3. Conservationists-Wisconsin-Biography.
I. Tanner, Thomas, 1936- .
II. Soil and Water Conservation Society (U.S.)
QH31.L618A68   1995
333.7'2'092—dc20
[B]   95-22547
                                            CIP

Aldo Leopold
The Man and His Legacy

Aldo planting Norway pines at the Shack gate, 1938

Contributors

CRAIG W. ALLIN
Professor and Chairman, Department of Politics, Cornell College,
Mount Vernon, Iowa

BRUCE BABBITT
Governor of Arizona, Phoenix

CHARLES C. BRADLEY
Director of Research, Leopold Memorial Reserve,
and Visiting Professor, University of Wisconsin,
Baraboo, Wisconsin

NINA LEOPOLD BRADLEY
Coordinator of Research, Management, and Educational Projects;
Leopold Memorial Reserve and Sand County Foundation;
Baraboo, Wisconsin

J. BAIRD CALLICOTT
Professor of Philosophy and Natural Resources,
University of Wisconsin, Stevens Point

RAYMOND F. DASMANN
Professor of Ecology and Member, Environmental Studies Board,
University of California, Santa Cruz

SUSAN FLADER
Professor of History, University of Missouri,
Columbia

HUEY D. JOHNSON
Director, The New Renaissance Institute, Sausalito, California

CONTRIBUTORS

SHARON KAUFMAN
Naturalist, Clinton County Conservation Board,
Grand Mound, Iowa

CARL LEOPOLD
William C. Crocker Distinguished Scientist,
Boyce Thompson Institute for Plant Research, Cornell University,
Ithaca, New York

ESTELLA LEOPOLD
Professor of Botany and Forest Resources,
University of Washington, Seattle

FREDERIC LEOPOLD
Manufacturer and Wildlife Researcher,
Burlington, Iowa

LUNA LEOPOLD
Professor Emeritus of Geology, University of California,
Berkeley

DALE McCULLOUGH
A. Starker Leopold Professor of Wildlife Management,
Department of Forestry and Resource Management,
University of California, Berkeley

CURT MEINE
Ph.D. Candidate in Land Resources, University of Wisconsin,
Madison

RODERICK NASH
Professor of History and Environmental Studies,
University of California, Santa Barbara

Contents

CONTENTS

Foreword

Twenty-five years have passed since these essays and reflections on Aldo Leopold were first published—a rather remarkable fact to absorb for those of us who contributed to this volume and the gathering that precipitated it. Twenty-five times around the sun. One human generation. Four or five sandhill crane generations. One hundred monarch butterfly generations. Eighteen thousand high tides and eighteen thousand low tides. Forty-five more parts per million of atmospheric carbon dioxide. Two billion more people.

The earth, its soils and waters, all its diverse expressions of life, and all our interdependent lives upon it spin on together. I cannot help but recall a line from one of Leopold's early manuscripts: "...the privilege of possessing the earth entails the responsibility of passing it on, the better for our use, not only to immediate posterity, but to the Unknown Future, the nature of which is not given us to know."

I wonder, had Leopold revised that passage, if he might have shifted in his own sense of just who possesses what. I wonder if he might have nodded in assent to Robert Frost's opening line in "The Gift Outright:" "The land was ours before we were the land's." Frost and Leopold could have carried on quite a conversation. Frost, later in the same poem, said,

> Something we were withholding made us weak
> Until we found out that it was ourselves
> We were withholding from our land of living,
> And forthwith found salvation in surrender.

Leopold, in his foreword to *A Sand County Almanac*, wrote,

> We abuse land because we regard it as a commodity belonging to us. When we see land as a community to which we belong, we may begin to use it with love and respect.

And both might have appreciated the words of Kiowa poet and writer N. Scott Momaday from his essay "A First American Views His Land:"

> As an Indian I think: "You say that I *use* the land, and I reply, yes, it is true; but it is not the first truth. The first truth is that I *love* the land; I see that it is beautiful; I delight in it; I am alive in it."

Over these last twenty-five years, much has changed in this world that we are alive in; in our culture, politics, and economy; and in the ways we conservationists go about our work and our relationships. When we gathered in Iowa back then, the Midwestern farm crisis was rippling painfully over the surrounding countryside; the new compound word "biodiversity" had just been coined; the endangered northern spotted owl was becoming a symbol and scapegoat; a deep drought would soon come to the Midwest and sweeping fires to Yellowstone; and climate change was a matter of gathering data and concern and consensus, not of automatic ideological assault. We were wrestling with conservation challenges old and new, pausing to celebrate the positive accomplishments that Leopold inspired, even as we sought fresh pathways beyond the environmentalism of the time.

To read these essays now is to see that, however much has indeed changed, the core themes—in Leopold's experience and in our collective efforts to be better members of the land community—endure: the varied lessons we derive from ecology, the diverse values we discover in the wild, our need for a new and restorative agriculture, and the challenge of making the land ethic meaningful in an increasingly complex political and economic world that is also increasingly divorced from the land. To read the Leopold family's personal recollections of Aldo is to be reminded that, however great and abstract these themes seem, we explore them through the very real places and people with whom we share our lives. And how we go about our explorations reveals who we are. "Always Dad tried to be persuasive," daughter Nina recalls. "He had a very basic respect for other people's thoughts and never lashed out."

Now, twenty-five years on in the life of this particular collection, we turn our eyes forward again. We wonder what the next quarter century will bring. We wonder what the next generation will face and how we can best prepare them. So, for the younger readers of this book: I hope these essays will give you a sense of continuity and of foundations. I hope they won't seem too dated—and if they occasionally do, that that in itself will demonstrate that you are the inheritors of an ongoing, multigenerational effort to build a durable land ethic. You are the next to forge on into "the Unknown Future," to become the land's, to follow the first truth: to love the land, to see that it is beautiful, to delight in it, to be alive in it… along with the cranes, and the monarchs, and the tides.

Curt Meine

Preface

Most of the papers in this volume were first presented at the Aldo Leopold Centennial Celebration held at Iowa State University in October 1986. This book is not a proceedings of that event, however, because some papers have been added, others deleted or revised, and references to the celebration have been pared away. The result is a collection of new essays that honor a significant human life. In these papers we meet Aldo Leopold as scientist, activist, and philosopher; we meet him as a youth, eager young professional, and mature scholar. In these pages we are allowed to stand upon his shoulders and look to the future. But before we do, some words about our distinguished authors and others who made this volume possible, followed by a brief description of the book's structure and the intent of its several parts.

Although the book is not the proceedings of an event, it exists because of that event; the history of the Leopold celebration is the history of this volume. Both had their genesis in the autumn of 1983, when a lovely new film, "Aldo Leopold: His Life and Thought," reminded me that Leopold was born in Iowa in 1887 and raised in that state. As a professor of environmental studies at Iowa State University, I resolved immediately that the centennial of his birth should be appropriately observed in his home state. That resolve was transformed into action in the autumn of 1984, when a committee of five administrators and faculty members began planning the celebration. We were determined that the project should be worthy of

the man, and so one of our first tasks was to secure commitments
from outstanding speakers who would at once bring to our podium
distinction, scholarship, and the proven ability to communicate
with a general audience. In this we succeeded; our roster of speakers
reads like an Aldo Leopold all-star team: Ray Dasmann, one of the
world's most respected wildlife ecologists, whose textbook *Environmental Conservation* was dedicated to Leopold. Susan Flader, Leopold's biographer. Rod Nash, perhaps America's best-known environmental historian. Bruce Babbitt, governor of Arizona; we felt
that he, among all the 50 governors, stood on a record that most
nearly illustrated Leopold's land ethic at work in the world of real-
politik. Craig Allin, whose 1982 volume *The Politics of Wilderness
Preservation* had contributed to political history what Nash's
Wilderness and the American Mind had brought to the history of
ideas 15 years earlier. Huey Johnson, whose first act as chief of
California's vast resources agency was to tell his employees to read
Leopold's *A Sand County Almanac,* and who thereafter initiated
scores of projects that have survived the subsequent administration
while improving renewable resources, increasing employment, and
even turning a profit.

And more. The philosopher Baird Callicott, whose *Companion to
A Sand County Almanac* will be published in the centennial year.
Dale McCullough, whose studies of deer in Michigan and California
remind us that this animal and its predators had taught Leopold at
last to "think like a mountain." Charles Bradley, director of research
at the Leopold Memorial Reserve. Younger scholars, Curt Meine
and Sharon Kaufman, whose recent research on Leopold won them
places on our program among the aforementioned luminaries.
Balladeer Doug Wood, whose moving performance at The Maintenance Shop would include the premiere of a song about Leopold.

And still more. Aldo Leopold's four children, Luna, Nina, Carl,
and Estella, who built their own distinguished careers as scientists
and conservationists, as did their late brother Starker. They accepted our invitation to present seminars in their research specialties.
And finally, their Uncle Frederic, Aldo's younger brother. This remarkable man had stayed home to tend the family manufacturing
firm while Aldo followed his star to Yale, the Southwest, and Wisconsin. But the love of wildlife that their father had kindled in both
boys could be suppressed in neither. Fred built nesting boxes in his
yard, and his observations over four decades made him the world

authority on the nesting behavior of the wood duck, with many discoveries to his credit. (Aldo Leopold praised world-class amateur wildlife researchers in his essay, "Wildlife in American Culture." Had he been writing 30 years later, his brother could have been one of his examples.) So, this 91-year-old man brought his slides to Ames, kept a full-house audience enthralled for an hour, and then received their standing ovation.

I beg the reader's indulgence as I note briefly that the celebration was an unqualified success. Every speaker lived up to our high expectations, and all sessions flowed along without visible problems. In addition to our local audience, visitors came from throughout Iowa and adjacent states and from as far away as both coasts, just for the celebration. Attendance at the 19 sessions was 3,460. Associated activities included exhibits of wildlife art, Frederic's papers, and books by Leopold and our speakers; the reading of excerpts from *A Sand County Almanac* on public radio by ISU's extraordinary radio actor, Doug Brown; and—one month later—a band concert featuring "Eulogy for the Wilderness," by ISU composer Jeffrey Prater, who based this noble work upon a passage from *Almanac*.

Many persons and groups are due credit for making the celebration and thus this book possible. Cash contributions to the celebration were received from no fewer than 40 sponsors, including the ISU Committee on Lectures (funded by the Government of the Student Body), ISU Achievement Foundation, Iowa Humanities Board, Iowa Department of Natural Resources, a newspaper (*The Hawk Eye* of Burlington, Iowa), the ISU chapter of the science honorary Sigma Xi, several individuals, and some 24 departments, institutes, programs, colleges, councils, and student organizations at ISU. Generous gifts from Ducks Unlimited, Izaak Walton League of America, National Audubon Society, National Wildlife Federation, The Nature Conservancy, Society of American Foresters, and Soil Conservation Society of America confirmed the national rather than merely regional significance of the event.

The committee that spent two years planning the celebration included Robert Krotz, ISU's acting director of development; David Lendt, director of information; and Robert Summerfelt, professor and former department chairperson of Animal Ecology. Journalism professor Thomas Emmerson served for one year, to be succeeded in 1985-1986 by Stephen McDonnell, assistant dean for student organizations and activities. I am acting coordinator of the Environmental

Studies Program and served as the committee's chairperson.

George C. Christensen, vice-president for academic affairs, gave early and continuing support to the project. Thanks are due also to artist David Sauke, ISU's Animal Ecology Department, several student organizations, the local chapter of the Audubon Society, and others who, we regret, cannot possibly all be acknowledged by name.

Many of the kudos we've received have addressed the apparent smoothness with which the event occurred. Much credit for this is due Sandra Brooks, the committee's administrative assistant. Her ability to make things look easy while in fact solving several logistics problems simultaneously and graciously was a foil to my own tendency toward instant panic. Moreover, her work in making advance preparations was distinguished by thoroughness, initiative, and a pleasant demeanor toward all.

I am grateful for the assistance and especially the patience of my wife Sally and our son Tom. In his wisdom, Aldo Leopold shared many of his sand farm hours with his family. I did far less well while working on the celebration.

That the celebration should have produced a book is due to the entrepreneurial courage of the Soil Conservation Society of America and the initiative, industry, and patience of its editor, Max Schnepf. Finally, my thanks to Stewart Udall. Two decades ago and more, I admired his work as secretary of the interior and author of *The Quiet Crisis*, and knew of his own high regard for Leopold. Udall's dedication, distinction, and defense of the land have not diminished; thus my pleasure when he agreed to write our foreword.

* * *

The book is arranged into three parts. In the first, Leopold is the object of the authors' research. In the second, he is their mentor and inspiration. In the third, he is their kinsman.

The authors of Part I are Flader, Allin, Meine, Nash, and Callicott. These scholars document the evolution of Leopold's land ethic and its place in the history of American thought. They explain his contribution to wilderness preservation and his ideas about the interface of agriculture and conservation. Their perspective is that of the humanities and social sciences, specifically history, political science, literature, and philosophy, as enriched by their own love of things natural, wild, and free. (I note, incidentally, that four of these five will have books about Leopold or his ideas published or reissued during the centennial year.)

In Part II we move away from studies of Leopold to somewhat shorter and less formal papers by four persons who stand upon his shoulders in their own work. These four—Babbitt, Johnson, Mc-Cullough, and Dasmann—all have academic backgrounds in natural science or wildlife management, but their interests have taken them on to careers in public service, private-sector activism, and wildlife research, as well as the education of resource professionals, popularization of environmental concern, and intellectual leadership in the new environmentalism. The debt of all these writers to Leopold is made clear in their essays, especially in the intensely personal account that we asked Huey Johnson to prepare.

The focus of Part III is the Leopold family, both as subjects and authors. Leopold's brother takes us on family vacations in a nostalgically distant era; the impact of the northern forest on Aldo's life fairly shouts from between the lines. A generation later, weekends at the Shack work their influence upon his children. Later still, the Shack becomes part of the Leopold Memorial Reserve, a venture in restoration, preservation, and research. An interview with the Leopolds concludes the book.

* * *

A few months after Aldo Leopold's untimely death in 1948, a lengthy appreciation of him was published in *The Journal of Wildlife Management*. The author was Iowa State University's distinguished naturalist, Paul Errington, who wrote, "Let no one do him the disservice of fostering Leopoldian legends or Leopoldian dogmas. Knowing him as I have, I can say that he would not wish them to arise from his having lived.... I can imagine his gentle scorn at the thought of anything like elaborate statuary in his memory while despoilation and wastage of the land and its biota continue as usual."

With Errington's admonition in mind, this little volume is offered in the hope that it does not constitute verbal statuary and will do nothing to encourage legend and dogma. If instead it plays some small part in nudging the land ethic from inspired idea to cultural reality, it will have served its purpose.

Thomas Tanner

January 1987

I

The Man
and His Legacy

Early Forest Service days

1

Aldo Leopold and the Evolution of a Land Ethic

Susan Flader

The last two decades have witnessed an explosion of interest in the ethical basis for people's relationship with their environment, as conservationists, scientists, political leaders, and philosophers have sought an undergirding for what some feared might otherwise be an ephemeral flurry of public concern about environmental quality. In the resulting literature on environmental philosophy, one name and one idea recur more frequently than any others: Aldo Leopold and his concept of a land ethic.

Leopold articulated his environmental philosophy most powerfully in "The Land Ethic," capstone of *A Sand County Almanac,* the slim volume of natural history essays for which he is best known today. But "The Land Ethic" and the shorter, lighter, more illustrative vignettes that help illuminate it draw their clarity, strength, and enduring value from a lifetime of observation, experience, and reflection.

At his death in 1948, Leopold was perhaps best known as a leader—*the* leader—of the profession of wildlife management in America. He was also a forester and is regarded as the father of the national forest wilderness system. In tracing the evolution of his land ethic philosophy, however, we must look beyond his interest in wildlife and wilderness to his lifelong effort to understand the functioning of land as a dynamic system, a community of which we are all members. Though his ethical philosophy was an outgrowth of his entire life experience, this quest for its origins focuses in particular

3

on an aspect of his thought that has been less widely known, his concern about mountain watersheds and the problem of soil erosion.

Foundations

Aldo Leopold was born in 1887 in Burlington, Iowa, in his grandparents' home on a bluff overlooking the Mississippi River.[1] "Lugins-land"—look to the land—the family called the place, reflecting their German heritage and love of nature. The patriarch was Charles Starker, trained in Germany as a landscape architect and engineer, who instilled his keen esthetic sense in his daughter Clara, Aldo's mother. Clara had a pervasive influence on Aldo, eldest and admittedly the favorite of her four children, nurturing the esthetic sensitivity that would be so integral to his land ethic philosophy. Carl Leopold, Aldo's father, also had a formative influence on him. He was a pioneer in the ethics of sportsmanship at a time when the very notion of sportsman was taking shape in the United States.

Leopold frequently hunted with his father and brothers on weekends and arranged his classes so he would have weekday mornings free to cross the Mississippi for ducks or tramp the upland woods on the Iowa side. Summers he spent with the family at Les Cheneaux Islands in Lake Huron, fishing, hunting, and exploring. Despite the temptations of his avocation, he secured a remarkably good education in the Burlington schools, especially in English and history. Then, though he might have been expected to follow his father into management of the Rand & Leopold Desk Company, he went east to Lawrenceville Preparatory School in New Jersey, the Sheffield Scientific School at Yale, and eventually the Yale Forest School to prepare himself for a career in the new profession of forestry.

When he left for Lawrenceville in January 1904, it was with his mother's admonition to write frequently and "tell me everything." He wrote her from the train while passing through the mountains of Pennsylvania. He wrote again upon arriving in Lawrenceville, describing the lay of the land—"flat, but not so bad as I thought"—and yet again the next day, by which time he had taken an afternoon's tramp of some 15 miles and could pronounce himself "more

---

[1]Biographical details and generalizations are based primarily on the Aldo Leopold Papers in the University of Wisconsin Division of Archives and interviews with family members and friends of Aldo Leopold. All quotations from letters and unpublished manuscripts are from items in the Aldo Leopold Papers.

than pleased with the country." That letter ran four pages with detailed accounts of timber species, land use practices, and birds, as well as of the dormitory, his class schedule, problems in algebra and prospects in German, everything. From then on for a decade, through his school years in the East and his early career in the Southwest until after his marriage, Leopold would write at least weekly to his mother, less frequently to his father, brothers, and sister. By rough count he penned something on the order of 10,000 pages of letters. He had learned grammar and sentence structure in the Burlington schools, but he learned to write by writing. His mother insisted on it.

At Yale, Leopold's letters grew longer with more elaborate narrative and keen observation, the result of continued encouragement from home and more frequent and extended tramps through the countryside. He frequented special places—Juniper Hill, Marvelwood, Diogenes Delight, the Queer Valley—and wrote in detail of his encounters with foxes, deer, all manner of birds, moods, and weather. But when he began technical studies in forestry—dendrology, mensuration, silviculture, forest economics—the forays became less frequent as his dedication to the scientific challenges of the new profession absorbed his attention.

Leopold as Forester

Yale had established the first graduate school of forestry in the United States in 1900 with an endowment from the family of the nation's leading forester, Gifford Pinchot. The school promoted Pinchot's doctrine of scientific resource management and what Samuel Hays has characterized as the Progressive Era's "gospel of efficiency" (2). When Leopold graduated with a master of forestry degree in 1909 and went to work for Pinchot in the U.S. Forest Service, he was one of an elite corps of scientifically trained professionals who would develop administrative policies and techniques for the fledgling agency charged since 1905 with responsibility for managing the national forests. He was assigned to the new Southwestern District embracing Arizona and New Mexico territories.

His private reflections, expressed in letters home and recalled in essays written some three decades later, indicate that he was thoroughly enamoured with the stark beauty of the country. But his first publications, which soon began appearing in local Forest Service periodicals, reveal the extent to which he bought into the utilitarian emphasis of the forestry enterprise. The earliest was a bit of doggerel titled "The Busy Season" (3), which he inserted anonymously into a newsletter he edited for the Carson National Forest in New Mexico:

> There's many a crooked, rocky trail,
>   That we'd all like straight and free,
> There's many a mile of forest aisle,
>   Where a fire-sign ought to be.
>
> There's many a pine tree on the hills,
>   In sooth, they are tall and straight,
> But what we want to know is this,
>   What will they estimate?
>
> There's many a cow-brute on the range,
>   And her life is wild and free,
> But can she look at you and say,
>   She's paid the grazing fee?
>
> All this and more, it's up to us,
>   And say, boys, can we do it?
> I have but just three words to say,
>   And they are these: "Take to it."

As he was coming to grips with the myriad management problems of the Carson, Leopold was also struggling with a personal dilemma—how to win the attention of a Spanish senorita he had just met, when he was "1,000,000 miles from Santa Fe" and his rival was right on the scene. He called on his best resource, his skill at expressing himself in letters.

> My dear Estella—
>   This night is so wonderful that it almost hurts. I wonder if you are seeing the myriads of little "Scharfchens volken" I told you about—do you remember the "little sheep-clouds"?—I have never seen them so perfect as they are tonight. I would like to be out in *our Canon*—I don't know how to spell it so you will have to let me call it that—and see the wild Clematis in the moonlight—wouldn't you? I wonder if we could find four more little bluebells for you to pin at your throat—they were beautiful that evening in the dim twilight as they changed with the darkness into a paler and more unearthly blue—and finally into that color which one does not see but *knows*—simply because they are Bluebells.

After six months of extraordinary letters, Estella decided on Aldo; a year later the two were married. For the rest of his life, Estella would inspire and respond to Leopold's special esthetic sensitivity, as his mother had earlier, keeping it alive even during the long years when he was otherwise preoccupied with more worldly, practical affairs. Six months after their marriage, Leopold was off settling a range dispute in a remote area when he got caught in a flood and blizzard. He became a victim of acute nephritis, a kidney ailment that nearly took his life. After 18 months of recuperation and long hours of reflection on how he might live whatever time remained to him, he was able to return to light office work at the district headquarters in Albuquerque.

It was at this juncture that Leopold became involved in wildlife conservation, developing a new emphasis on cooperative game management that became a model for Forest Service activity around the nation. Though it was a line of work of his own choosing, an outgrowth of his early avocation, he approached it not in an esthetic mode but in the scientific, utilitarian spirit of the Forest Service. He would promote game management as a science, modeled on the principles and techniques of forestry. Game could bring nearly as much income to the region as timber or grazing uses of the forests, he calculated, if enough effort, intelligence, and money were committed to develop the resource. Extermination of wolves, mountain lions, and other predatory species was a key element in his early program, one he would live to regret (1).

By 1917 he had achieved national recognition for his successes in the Southwest and was beginning to publish his ideas on game management and forestry regularly in periodicals of nationwide circulation. From then until his death, he published frequently—never fewer than two articles a year, often more than a dozen. But for the next two decades his work was decidedly management-oriented. It was not until the last decade of his life that he began publishing the literary and philosophical essays for which he is best known today.

As one examines the contours of Leopold's life, it is ironic to find that the esthetic appreciation for wildlife that was so integral to his youth, so evident in his early letters, and so vital to his mature philosophic reflection was seemingly suppressed at mid-career, at least in his public persona, as he sought to develop a science and profession of wildlife management. The irony is compounded when one notes the extent to which he was pushing beyond traditional modes of thought in his understanding of the dynamics of southwestern watersheds by the early 1920s, developing an interpretation of the functional interrelatedness of virtually all elements of the system save wildlife. It was as if his effort to achieve parity for game animals within the Forest Service model of professional management limited his ken at the same time that he felt less constrained about challenging orthodoxy on larger issues. Thus, it is to his thinking about watersheds and soil erosion that we must turn if we would understand the evolution of his concept of a land ethic — his capacity to think about the system as a whole.

Southwestern Watersheds and Moral Obligation

Ever observant, Leopold had noted the condition of watersheds on national forests in the Southwest from his earliest days as a forest assistant. His first assignment had been to map and cruise timber along the route of a proposed wagon road that would have to clamber high over the Blue Range on the Apache Forest in eastern Arizona because erosion had foreclosed all possibility of a more logical route through the once-lush bottomlands of the Blue River. When he was promoted in 1919 to assistant district forester and chief of operations in charge of business organization, personnel, finance, roads and trails, and fire control on 20 million acres of national forests in the Southwest, Leopold had an opportunity to observe conditions anew as he crisscrossed the forests on inspection trips. By then erosion had washed out nearly 90 percent of the arable land along the Blue. Of 30 mountain valleys he tallied in southwestern forests, 27 were already damaged or ruined.

Leopold thought long and hard about soil erosion. He wrote about the problem in his inspection reports, trying to make foresters in the field more cognizant of changes occurring before their eyes. He spoke about it to the New Mexico Association for Science in a strongly worded warning, "Erosion as a Menace to the Social and Economic Future of the Southwest," and reached for national attention to the problem through articles in the *Journal of Forestry*. He even grappled with the philosophical meanings of the phenomenon in a remarkable essay, "Some Fundamentals of Conservation in the Southwest," that was found in his desk after his death.

In "Some Fundamentals," written in 1923, Leopold probed the causes of the degradation that was reducing the carrying capacity of southwestern ranges and considered the implications for human ethical behavior. "The very first thing to know about causes," he wrote, "is whether we are dealing with an 'act of God,' or merely with the consequences of unwise use by man" (*10*). Through analysis of evidence from tree rings, archaeology, and history, he concluded that the deterioration of organic resources in the Southwest could not be attributed simply to climatic change (an act of God). But the nature of the climate, characterized by periodic drouth, had resulted in a delicately balanced equilibrium that was easily upset by man. Overgrazing, resulting from overstocking without regard to recurring drouth, was the outstanding factor in upsetting the equilibrium, in his view. This conclusion that human beings bore responsibility for unwise land use led him to a philosophical discussion under the subtitle "Conservation as a Moral Issue."

He began with an admonition from Ezekiel:

> Seemeth it a small thing unto you to have fed upon good pasture, but ye must tread down with your feet the residue of your pasture? And to have drunk of the clear waters, but ye must foul the residue with your feet?

Ezekiel seemed to scorn poor land use as something damaging "to the self-respect of the craft and society" of which one was a member. It was even possible, Leopold thought, that Ezekiel respected the soil "not only as a craftsman respects his material, but as a moral being respects a living thing." Leopold found support for his own intuitive feeling that there existed between man and earth a deeper relation than would follow from a mechanistic conception of nature in the organicism of the Russian philosopher, P. D. Ouspensky, who regarded the whole earth and all its parts as possessed of soul or consciousness. "Possibly, in our intuitive perceptions, which may be truer than our science and less impeded by words than our philosophies," Leopold suggested, "we realize the indivisibility of the earth — its soil, mountains, rivers, forests, climate, plants, and animals, and respect it collectively not only as a useful servant but as a living being."

Realizing that this premise of a living earth might be too intangible for many people to accept as a guide to moral conduct, Leopold launched into yet another philosophic issue: "Was the earth made for man's use or has man merely the privilege of temporarily

possessing an earth made for other and inscrutable purposes?" Because he recognized that most people were heir to the mechanistic,
anthropocentric scientific tradition or—like his wife, a devout
Roman Catholic—professed one of the anthropocentric religions, he
decided not to dispute the point. But he couldn't resist an observation: "It just occurs to me, however, in answer to the scientists, that
God started his show a good many million years before he had any
men for audience—a sad waste of both actors and music—and in
answer to both, that it is just barely possible that God himself likes to
hear birds sing and see flowers grow." Even granting that the earth
is for man, there was still a question: "What man?" Four cultures
had flourished in the Southwest without degrading it. What would
be said about the present one?

> If there be, indeed, a special nobility inherent in the human race—
> a special cosmic value, distinctive from and superior to all other life—
> by what token shall it be manifest? By a society decently respectful of
> its own and all other life, capable of inhabiting the earth without de-
> filing it? Or by a society like that of John Burroughs' potato bug,
> which exterminated the potato, and thereby exterminated itself?

We can only guess why Leopold decided not to publish "Some
Fundamentals." One reason might have been a certain discomfort
with the inconclusiveness of his philosophical arguments. Another
might have had to do with criticism of his interpretation of the
causes of soil erosion.

One colleague to whom he sent the draft for review, Morton M.
Cheney of the lands division, took him to task for overemphasizing
the destructiveness of erosion. Like most other foresters and scientists of the time, Cheney viewed erosion as a natural, ongoing geo-
logic process, a "world-building factor" that would ultimately
smooth the rough uplands of the Southwest and create an immense
area of agricultural land. In retrospect, it is clear that Cheney was
explaining to Leopold the classic geomorphological theory of landscape development of William Morris Davis, who described the
stages through which mountains are uplifted and erode to pene-
plains. Davis's "cycles of erosion" had much in common with the
prevailing model of forest ecology, the stages of plant succession to
climax, which underlay forest policy in the Southwest. Both were
developmental models, defining predictable stages leading to a
stable endpoint or equilibrium that would persist indefinitely, unless

assaulted by some force acting from outside the system. Thus Cheney articulated essentially the Forest Service prescription for management: protect the climax forests at the headwaters from fire, which will retard some erosion, and don't worry too much about erosion downstream, a natural process in any case.

Leopold didn't need a disquisition on erosion from Cheney. He had undoubtedly studied all that at Yale. But he knew it did not explain what he observed on the ground. He must also have been enormously frustrated by the unwillingness of his colleagues to recognize any responsibility on the part of the Forest Service to deal with the problem. One effect of Cheney's critique must have been to make Leopold realize that he was going to have to provide a much more persuasive analysis of the causes of erosion, starting from facts anyone could see. This he accomplished a year later in an extraordinary piece of observation and inferential reasoning dealing with the relationship between grazing, fire, plant succession, and erosion. "Grass, Brush, Timber and Fire in Southern Arizona" still stands as a landmark in ecological literature. In it he issued a direct challenge to Forest Service dogma: "Fifteen years of forest administration were based on an incorrect interpretation of ecological facts and were, therefore, in part misdirected" (5).

Leopold's theory, in brief, held that southwestern watersheds had maintained their integrity despite centuries of periodic wildfire set by lightning or Indians. Overgrazing by cattle beginning in the 1880s thinned out the grass needed to carry fire, and brush species, now free of both root competition and fire damage, began to "take the country," thus further reducing the carrying capacity for cattle. By the time brush and, higher up, yellow pine had grown dense enough to carry fire, the Forest Service had arrived on the scene to prevent it. Trampling by cattle along watercourses allowed erosion to start, and the grass was no longer sufficient to prevent devastation. Contrary to Forest Service doctrine, Leopold found evidence that fire, far from being an unmitigated evil, was natural and even beneficial, and that grass was a much better conserver of watersheds than were trees or brush. While the Forest Service was willing to acquiesce in some overgrazing and erosion in order to reduce the fire hazard, Leopold was willing to take an added risk of fire in order to maintain the integrity of the watersheds. The Forest Service was thinking of the commodity values of cattle and timber. Leopold was thinking of the whole system.

Although he still had not broken completely out of the mold in which he had been educated and there was much research yet to come, especially on such matters as fire ecology, climate change, and the role of wildlife, Leopold finally had a theory to explain the degradation of southwestern watersheds and was able to consider its implications for administrative policy and for human ethical behavior. Even as he was working out the new theory, he also prepared a comprehensive "Watershed Handbook" for the district, to train foresters to analyze and treat erosion problems, and he took his case for conservative land use to the public in an article, "Pioneers and Gullies," in *Sunset Magazine*. In addition to posing several short- and medium-range policy options — artificial works, better regulated grazing permits, a leasing system — he also addressed the long-range ethical issue, for the first time in print. But it was just a short prediction: "The day will come when the ownership of land will carry with it the obligation to so use and protect it with respect to erosion that it is not a menace to other landowners and to the public" (4). Frustrated as he was by the failure of the Forest Service to address the erosion problem, Leopold realized that the issue ultimately would have to be joined on private lands. Hence the need for a sense of ethical obligation — supported, he hoped, by a legal concept of contingent possession or some other public policy.

It would be another decade before Leopold would publish anything substantial on conservation as a moral issue. But when he did, it would relate integrally to his deepened ecological understanding of southwestern watersheds. Once he could explain in sufficient ecological detail the phenomenon of erosion, or other aspects of the dynamic functioning of ecosystems through time, he would no longer need to call on the authority and terminology of the philosophers. He could stand secure on his own ground. Was the earth living or not? Was the earth primarily for man's use or not? Definitive philosophical answers to such questions no longer seemed critical as the basis for an ethic. It was more important to grasp the interrelatedness of all elements of the system, physical and biological, natural and cultural, and to appreciate the extent to which human as well as biotic interests were dependent upon maintenance of the integrity of that system through time.

In 1924 Leopold left the Southwest for Madison, Wisconsin, to assume a new position as associate director of the Forest Products Laboratory. The laboratory at that time was the agency's principal research center. Though he was offered the position in part because of his intense scientific curiosity and interest in research as demonstrated in his work on southwestern watersheds, the offer did not imply wholesale acceptance of his ideas within the Forest Service. Nor did the new job offer a fully satisfying scope for his interests. As its name implied, the laboratory dealt more with research concerning the uses of trees than with the living forest.

Four years later, Leopold left the Forest Service to venture full-time into the relatively new field of game management, which had been his hobby long before it became his profession. Drawing on his experience inspecting and analyzing the condition of watersheds in southwestern forests, he began a series of game surveys in the north central states with funding from the Sporting Arms and Ammunition Manufacturers' Institute. The surveys augmented his understanding of interrelationships between wildlife and the land community and launched him on another major project, the writing of a comprehensive text for the new field. More or less unemployed during the darkest years of the depression, 1931 and 1932, while he was writing his now-classic *Game Management,* Leopold readily accepted a temporary position with Franklin Roosevelt's New Deal in the spring of 1933 and headed back to the Southwest. His assignment was to supervise erosion control work by the newly established Civilian Conservation Corps in dozens of camps in the national forests.

He found the problems as evident as when he had left the region a decade earlier. The causes and processes of erosion were still not well understood within the Forest Service, and there had as yet been no changes in grazing regulations. The worst problems were on private land, giving urgency to the need Leopold had identified in "Pioneers and Gullies" for institutional incentives to conservative land use. Small wonder then that he drew heavily on watershed problems in the Southwest when, for the first time in public or in print, he developed the case for a land ethic. The occasion was the annual John Wesley Powell Lecture to the Southwestern Division of the American Association for the Advancement of Science.

The Conservation Ethic

"The Conservation Ethic" was by all odds the most important address of Leopold's career. It begins with a passage familiar to readers of his later "Land Ethic": "When god-like Odysseus returned from the wars in Troy, he hanged all on one rope some dozen slave-girls of his household whom he suspected of misbehavior during his absence" (6). Disposal of property was a matter of expediency. Leopold went on to make the case for the extension of ethical criteria to all human beings, to social relationships, and eventually to the land community; he asserted that this extension of ethics was actually a process in ecological evolution. An ethic, he said, "may be regarded as a mode of guidance for meeting ecological situations so new or intricate, or involving such deferred reactions, that the path of social expediency is not discernible to the average individual." Discussing the role of ecology in history and its relationship to economics and eventually to ethics, he returned repeatedly to the vexing problem of erosion.

Leopold could identify three possible motives for soil conservation: self-interest, legislation, and ethics. Self-interest did not pay, especially on the marginal private land most subject to abuse. Legislative efforts to regulate land use would force tax delinquency, especially during a depression, and public ownership could not possibly go far enough. "By all the accepted tenets of current economics and science we ought to say, 'let her wash'," Leopold concluded. Staple crops were overproduced, population was stabilizing, science was still raising yields, government was spending millions to retire lands, "and here is nature offering to do the same thing free of charge; why not let her do it?" This was economic reasoning. "*Yet no man has so spoken.*" To Leopold this fact was significant: "It means that the average citizen shares in some degree the intuitive and instantaneous contempt with which the conservationist would regard such an attitude." In this intuitive reaction Leopold saw the embryo of an ethic.

"The Conservation Ethic" is curiously transitional, much more economically based and management-oriented than his 1923 discussion of conservation as a moral issue and without appeals to Ouspensky or other philosophers. By the time the speech was published, Leopold had accepted a new chair of game management created for him in the Department of Agricultural Economics at the University of Wisconsin. There he would help devise new social and economic tools to deal with problems of land utilization and resource conservation. His emphasis in the address on ecologically based management by private owners on their own land was almost certainly in anticipation of his new position, where he would work not only with government agencies but also, and especially, with farmers. Even here, however, there is a mixture of older concepts with new: "It is no prediction but merely an assertion that the idea of controlled environment contains colors and brushes wherewith society may someday paint a new and possibly a better picture of itself." The "idea of controlled environment" — this confidence in the possibility of control, which permeates "The Conservation Ethic" and also *Game Management,* published the same year, is straight out of the Progressive Era conservation tradition of Gifford Pinchot. It assumes that scientific intelligence can learn enough about the system to exert complete control, an assumption that Leopold's invocation of an ecological attitude was even then beginning to challenge. The resolution would await yet another stage in his evolution of a land ethic.

In his new position at the university, he applied his concern about watershed integrity and soil erosion to southwestern Wisconsin. In a severely eroded watershed known as Coon Valley, Leopold and colleagues persuaded H. H. Bennett, chief of the federal Soil Erosion Service, to establish a pioneering demonstration on the integration of land uses — soil conservation, pasturage, crops, forestry, and wildlife — that would use trained technicians and hundreds of local farmers. Not long after the project began, Leopold expressed the difficulty and necessity of integrating land uses on private land in a wide-ranging critique of the single-track agencies and public-purchase panaceas of the New Deal. "Conservation Economics" is another classic Leopold essay, further substantiating the case for wise use of private as well as public land.

Philosophers speak of the need for a metaphysic, a theory of the scheme of things, as an undergirding for an ethic. Leopold no longer looked to philosophers for a metaphysic; he simply wanted to understand what was happening on the land. But he was still struggling to understand how the system functions and how human beings relate to it. In 1935 he returned again to the erosion puzzle in the Southwest, trying his hand at working out a theory that would be mutually acceptable to foresters, ecologists, geologists, and engineers concerning the role of man, and other factors such as climate and topography, in causing erosion. In order to counter an explanation of climate-induced synchronous timing of erosion episodes proposed by the eminent Harvard geologist, Kirk Bryan, Leopold wrote "The Erosion Cycle in the Southwest," in which he came up with an elaborate theory of random timing that allowed a role for human agency. As with his 1923 effort, he sent the manuscript for review, including a copy to Bryan. And again, after receiving Bryan's critique, he decided not to publish; instead, he urged his son Luna to study at Harvard with Bryan to see if he could solve the problem.[2]

Throughout the second half of the decade, Leopold continued his intellectual struggle for a better understanding of the system. He sensed the inadequacy of prevailing models without being able to put his finger on the problem. Sometimes, as ideas grow and change, a certain dramatic experience can trigger a rearrangement of elements, resulting in a new theory of the scheme of things. For Leopold the trigger might have been the juxtaposition of several key field experiences in the mid-1930s: a trip to study game management and forestry in Germany, during which he was appalled by the highly artificial system of management that created a host of unanticipated problems; the acquisition of his own sand country farm, where he began to experience first-hand the imponderables of even the best-intentioned management; and, perhaps most vital, two hunting trips to the Rio Gavilan in the Sierra Madre of northern Chihuahua.

The Sierra Madre—just south of the border from the Southwest Leopold had struggled so long to understand, but protected from overgrazing by Apache Indians, bandits, depression, and unstable administration—still retained the virgin stability of its soils and the integrity of its flora and fauna. The Gavilan River still ran clear between mossy, tree-lined banks. Fires burned periodically without any apparent damage, and deer thrived in the midst of their natural predators, wolves and mountain lions. "It was here," Leopold reflected years later, "that I first clearly realized that land is an organism, that all my life I had seen only sick land, whereas here was a biota still in perfect aboriginal health." The vital new idea for Leopold was the concept of biotic health. It was that idea that finally gave him a model, a way of conceptualizing the system, that could become the basis for his mature philosophy.

A Biotic View of Land

Leopold worked out his first comprehensive statement of the new scheme in a paper titled "A Biotic View of Land," which he deliv-

---

[2]World War II intervened and Aldo died before Luna was able to complete his Ph.D. dissertation on "The Erosion Problem of Southwestern United States" at Harvard in 1950.

ered in 1939 to a joint session of the Ecological Society of America and the Society of American Foresters. The biotic idea represented a shift from the older conservation idea of economic biology, with its emphasis on sustained production of resources or commodities, to a recognition that true sustained yield requires preservation of the health of the entire system. Scarcely five years earlier, Leopold himself had asserted that "the production of a shootable surplus is the acid test of the sufficiency of a conservation system." Now he was distinguishing between the old economic biology, which conceived of the biota as a system of competitions in which managers sought to give a competitive advantage to useful species, and the new ecology, which "lifts the veil from a biota so complex, so conditioned by interwoven cooperations and competitions, that no man can say where utility begins or ends" (7). Though he had clearly been moving toward such a view years earlier in his work on southwestern watersheds, the 1939 statement marked the first significant publication in which his thinking about wildlife was fully integrated in the new conception. Thus, better than anything else he wrote, "A Biotic View of Land" signaled the maturity of Leopold's thinking.

It was here that he first presented the image of land as an energy circuit, a biotic pyramid: "a fountain of energy flowing through a circuit of soils, plants, and animals." It was here too that he drew ecological interrelationships into an evolutionary context. The trend of evolution, he suggested, was to elaborate and diversify the biota, to add layer upon layer to the pyramid, link after link to the food chains of which it was composed. Leopold posited a relationship between the complex structure of the biota and the normal circulation of energy through it — between the evolution of ecological diversity and the capacity of the land system for readjustment or renewal, what he would come to term land health.

Biotas seemed to differ in their capacity to sustain conversions to human occupancy. Drawing once again on his understanding of the Southwest, Leopold contrasted the resilient biota of Western Europe, which had maintained the fertility of its soils and its capacity to adapt to alterations despite centuries of strain, with the semiarid regions of America, where the soil could no longer support a complex pyramid and a "cumulative process of wastage" had set in. The organism would recover, he explained, "but at a low level of complexity and human habitability." Hence his general deduction: "the less violent the man-made changes, the greater the probability of successful readjustment in the pyramid."

The biotic idea Leopold articulated, though deeply a product of his own thought and experience, was part of a larger conceptual reorientation in the biological sciences during the 1930s. These were the years, according to historians of science, when various strands of evolutionary and ecological theory, separated during the furor over Darwin's *Origin of Species* (1859), began to fuse into a broad unified theory. Among ecologists it has become known as the "ecosystem" concept, after a term suggested by the British ecologist A. G. Tansley in 1935.[3] The new conception postulated a single integrated system of material and energy. The system was not driven by any one factor, such as climate, and hence was not best understood as developing through predictable stages to a stable endpoint, as in the older models of forest succession and erosion that had troubled Leopold in the Southwest. Rather, it was in constant flux, subject to unpredictable perturbations that entailed a continual process of reciprocal action and adjustment. Agents of disturbance—whether fire, disease, grazing animals, predators, or man—which had been viewed by early forest ecologists as acting from outside the system to subvert the normal successional stages or the climax equilibrium, were now viewed as functional components within the system.

Not everyone who adopted the ecosystem model drew from it the same implications for land management or for ethical behavior. Human beings, an exceptionally powerful biotic factor, were now clearly located *within* the system; did that mean they could manipulate it to their own ends with moral impunity? Donald Worster, who has written about the history of ecological thought in *Nature's Economy*, has noted the emphasis in recent ecosystem analysis on concepts such as productivity, biomass, input, output, and efficiency, which he views as metaphors for the modern corporate, industrial system bent on economic optimization and control. He also views modern ecosystem analysis as reductionist, breaking down the living world into readily measurable components that retain no image of organism or community. He is, therefore, highly skeptical of the ecosystem concept as the basis for an environmental ethic (*11*).

A caveat is in order. Science alone is hardly an adequate basis for

[3]Leopold's biotic idea, with its image of the land pyramid, probably owes more to another British ecologist, Charles Elton, and to an American scientist, Walter P. Taylor, both personal friends. Leopold did not use the term "ecosystem" in his writing, but he used "ecology" or "biotic idea" to express the same concept.

an ethic, and Leopold realized as much. For all his commitment to understanding how the system functioned, even Leopold did not insist that science provided all the answers. What he gained from the biotic view was a new humility about the possibility of ever understanding the system fully enough to exercise complete control. As he expressed it in a speech on "Means and Ends in Wildlife Management," some managers had admitted their "inability to replace natural equilibria with artificial ones," and their unwillingness to do so even if they could. The objective of management, as he now viewed it, was to preserve or restore the capacity of the system for sustained functioning and self-renewal. He would do this by encouraging the greatest possible diversity and structural complexity and minimizing the violence of man-made changes. The techniques of management might remain much the same, but the ends (of wildlife management, at least) were now fundamentally altered. The ends, he realized, were a product of the heart as much as of the mind.

Leopold came to a deeper personal understanding and appreciation of the new biotic idea and its implications for land management through his own participation in the land community at his "shack" in central Wisconsin's cutover, plowed up, worn out, and eroded sand country. All during the late 1930s and into the 1940s, while he was struggling to put the biotic idea and the land health concept on paper, he was also struggling to rebuild a diverse, healthy, esthetically satisfying biota on his farm. His journals of the shack experience record a daily routine of planting and transplanting—wildflowers, prairie grasses, shrubs and trees, virtually every species known to be native to the area. But the journals also record his tribulations. Take pines, for example, of which the family planted thousands every year. The first year, 1936, more than 95 percent were killed by drouth within three months. Another year, rabbits attracted to brush shelters he had built for the birds trimmed three-quarters of the white pines in the vicinity. Other times the culprits were deer or rust or weevils or birds alighting on the candles or vandals cutting off the leaders or flood or fire. Fire could be discouraging, especially if set by a trespassing hunter, but it also brought new life—sumac and wild plum, blackberry, bluestem, poison ivy and, most exciting to Leopold, natural reproduction of jackpines from cones undoubtedly opened by the heat.

The shack experience engendered in Leopold a profound humility in his use of the manager's tools, as he became acutely aware of the innumerable, ofttimes inscrutable factors involved in life and death, growth and decay. It also led him to ponder the basis for the individual decisions he found himself making every day—whether to plant something as useless as a tamarack (yes, because it was nearly extinct in the area and it would sour the soil for lady's-slippers), what to do about the sandblow on the hill (leave it as testimony to history and also as a habitat for certain species like little Linaria that would grow only there), whether to favor the birch or the pine where the two were crowding each other (he loved all trees, but he was "in love" with pines). He realized that ethical and esthetic values could be a guide for individual decisions, not a substitute for them. And he also gained a sense of belonging to something greater than himself, a continuity with all life through time. This he expressed in a series of vignettes that ultimately found their way into *A Sand County Almanac.*

Toward a Land Ethic

Toward the end of the 1940s, Leopold tried again to make the case for a conservation ethic. This new effort, for a 1947 address to the Garden Club of America on "The Ecological Conscience," was less ambitious than his 1933 "Conservation Ethic." He drew on four issues in which he had been involved in Wisconsin, including a wrenching debate about the state's "excess deer" problem that had preoccupied him for years. The public thought only about conserving deer because they were unable to see the land as a whole. As in each of his previous efforts to make the case for a sense of obligation to the community going beyond economic self-interest, he also addressed the problem of soil erosion, this time analyzing the failure of soil conservation districts in Wisconsin to achieve anything beyond those few remedial practices that were immediately profitable to the individual farmer. From his unpublished papers we know that he had been trying for several years to articulate the land health concept in relation to land use and erosion, but he made no effort to do so in "The Ecological Conscience" (8). The speech was not particularly significant—except for one pregnant sentence setting forth the criteria of an ethic, which would ultimately find a more appropriate context in his most famous essay, "The Land Ethic."

Later that year and in early 1948, Leopold substantially reshaped and revised the collection of essays for which he had been seeking a publisher all decade. It would begin with a selection of vignettes from the shack, arranged by month in the style of an almanac. That would be followed by sketches recounting various episodes in his career that taught him the meaning of conservation. Then he would conclude with several meatier essays, culminating in a comprehensive statement of his ethical philosophy. "The Land Ethic" incorporates segments from three previous essays, all thoroughly reworked and integrated with new material to reflect his current thinking. From "The Conservation Ethic" he drew the notion of the ecological and social evolution of ethics and the role of ecology in history; from "A Biotic View of Land" the image of the land pyramid, of land as an energy circuit; and from "The Ecological Conscience" the case for obligations to land going beyond economic self-interest. His efforts of a decade to articulate the concept of land health and the relationships between economics, esthetics, and ethics, filling numerous handwritten, heavily interlined pages, finally found compelling expression. And the whole came to a focus in the most widely quoted lines in the entire Leopold corpus: "A thing is right when it tends to preserve the integrity, stability, and beauty of the biotic community. It is wrong when it tends otherwise" (9).

Integrity, stability, beauty: the fundamental criteria of the land ethic. Integrity, referring to the wholeness or diversity of the community: the precept to retain or restore, insofar as possible, all species still extant that evolved together in a particular biota. Stability, embodying the concept of land health: the precept to maintain or restore an adequately complex structure in the biotic pyramid, so that the community has the capacity for sustained functioning and self-renewal. Beauty, the motive power of the ethic: the precept to manage for values going beyond the merely economic — and, probably also, an allowance for the subjective tastes of the individual. The three tenets were interrelated. Elsewhere in his writings Leopold had referred to an assumed relationship between the diversity and stability of the biotic community as "the tacit evidence of evolution" and "an act of faith." And as early as 1938 he had posited a relationship among all three tenets, and utility as well, in an unpublished fragment titled "Economics, Philosophy and Land": "We may postulate that the most complex biota is the most beautiful. I think there is much evidence that it is also the most useful. Certainly it is the most permanent, i.e., durable. Hence there is little or no distinction between esthetics and utility in respect of biotic objective."

The three cardinal tenets of the land ethic, first voiced in his address, "The Ecological Conscience," can also be discerned, in somewhat different terminology, in Leopold's earlier formulations of his ethical philosophy in 1923 and 1933. This supports a conclusion that his mature expression involved a deeper understanding of the functioning of the land system and a more cogent articulation rather than a change in fundamental values. All these formulations, and "A Biotic View of Land" as well, were premised on a conception of land as an interrelated, indivisible whole, a system that deserved respect as a whole as well as in its parts. In 1923 Leopold drew on his intuitive perceptions, buttressed by the concepts and terminology of Ouspensky and other philosophers and poets. Later he would base his conception of the land community on the findings of ecology, but his willingness to trust to intuition remained. In 1933, for example, he wrote, "Ethics are possibly a kind of advanced social instinct in-the-making," revising the phrase in 1948 to "community instinct." Each formulation also emphasized the notion of *obligation* to the whole, rather than focusing on the *rights* of individual constituents, whether human, animal, vegetable, or mineral. Leopold did not deny that nonhuman entities had rights, and occasionally he even referred to a species' "biotic right" to existence, but he was too much concerned with securing acceptance of his major premises to risk alienating people by entering the thicket of the rights debate.[4]

Leopold articulated his ethical philosophy out of a profound conviction of the need for moral obligation in dealing with the dissolution of watersheds in the Southwest. Hence his emphasis on the integrity of the system. Each successive reformulation of his philosophy was stimulated at least in part by his continuing concern for the erosion problem and advances in his understanding of the ecological processes involved. Remarkably, it was his compelling concern and curiosity about the phenomenon of erosion, which was never a major professional responsibility, rather than his lifelong interest in wildlife, which became his profession, that led him to his conviction of the need for a land ethic and his understanding of the

---

[4]Because most philosophers are concerned with individuals rather than with communities and with rights rather than with obligations, Leopold's land ethic has often been distorted, disparaged, or dismissed in philosophical circles. Among those who have studied his work carefully enough to puzzle through some of the vagaries of language and seeming inconsistencies, Bryan Norton places Leopold in the tradition of American pragmatism and J. Baird Callicott identifies him as heir to the bio-social ethical tradition of David Hume and Charles Darwin. Both are reasonable in light of Leopold's education, reading, and experience.

biotic idea on which it was grounded. But once he had grasped the biotic concept, through which he finally integrated wildlife fully into his understanding of the functioning of the land community, it was his sensitivity to the esthetics of wildlife that would enable him to convey a sense of the land community and the land ethic to others. *A Sand County Almanac* is the case in point.

Leopold's fascination with the new biotic concept, especially the role of evolution, energy, and land health, led to an explosion of the classic literary essays for which he is best known today: "Marshland Elegy," with its haunting image of sandhill cranes, evolutionarily among the most ancient of species, standing in the peat bogs of central Wisconsin "on the sodden pages of their own history"; "Clandeboye," where the western grebe, also of ancient lineage, "wields the baton for the whole biota"; "Odyssey," the saga of two atoms cycling through healthy and abused systems; "Song of the Gavilan," where food is the continuum in the stream of life; "Guacamaja," a disquisition on the physics of beauty, recording the discovery of the numenon of the Sierra Madre, the thick-billed parrot; and "Thinking Like a Mountain," in which the wolf becomes metaphor for the functioning system.

These essays had a purpose with respect to Leopold's notion of the evolution of an ethic. Their purpose was to inspire respect and love for the land community, grounded in an understanding of its ecological functioning. Leopold would motivate that understanding of the whole by focusing the reader's attention on the subtle dramas inherent in the roles of wolf, crane, grebe, parrot, even atom, in the scheme of things. The essays were Leopold's attempt to develop a metaphysic, or an esthetic — to stimulate perception that might lead people to the transformation of values required for a land ethic. He would motivate not by inciting fear of ecological catastrophe or indignation about abused watersheds but rather by leading people from esthetic appreciation through ecological understanding to love and respect.

He had thus come full circle in his own development — from his youth, in which esthetic appreciation for wildlife provided his personal motivation to enter a career in conservation, through his professional experience in forestry and his concern about watersheds, which stimulated his consciousness of the need for an ethical obligation to land, to his maturity as an ecologist, when he successfully integrated all the strands of his previous experience. Reflecting on the process by which he himself had come to ecological and ethical consciousness, he would now inspire others along a similar route.

REFERENCES

1. Flader, Susan L. 1974. *Thinking like a mountain: Aldo Leopold and the evolution of an ecological attitude toward deer, wolves, and forests.* University of Missouri Press, Columbia, Missouri.
2. Hays, Samuel P. 1959. *Conservation and the gospel of efficiency: The progressive conservation movement, 1890-1920.* Harvard University Press, Cambridge, Massachusetts.
3. Leopold, Aldo. 1911. *The busy season.* The Carson [National Forest] Pine Cone (July).
4. Leopold, Aldo. 1924. *Pioneers and gullies.* Sunset Magazine 52 (May): 15-16, 91-95.
5. Leopold, Aldo. 1924. *Grass, brush, timber, and fire in southern Arizona.* Journal of Forestry 22 (October): 1-10.
6. Leopold, Aldo. 1933. *The conservation ethic.* Journal of Forestry 31 (October): 634-643.
7. Leopold, Aldo. 1939. *A biotic view of land.* Journal of Forestry 37 (September): 727-730.
8. Leopold, Aldo. 1947. *The ecological conscience.* The Bulletin of the Garden Club of America (September): 45-53.
9. Leopold, Aldo. 1949. *A sand county almanac and sketches here and there.* Oxford University Press, New York, New York.
10. Leopold, Aldo. 1979. *Some fundamentals of conservation in the Southwest.* Environmental Ethics 1 (Summer): 131-141.
11. Worster, Donald. 1977. *Nature's economy: The roots of ecology.* Sierra Club Books, San Francisco, California.

2

The Leopold Legacy
and American Wilderness

Craig W. Allin

In the spring of 1980 I was completing a book about wilderness preservation, about the evolution of public lands policy, from a policy of unrestrained development and hostility toward wilderness to one of modestly restrained development and at least lip service to the values of wilderness preservation (2). I wanted a quotation with which to begin, noble words that would cut to the heart of the issue, words that could serve as a kind of scripture for the volume that followed. The words I chose were Aldo Leopold's words. They appear in the preface to *A Sand County Almanac* (20):

> Like winds and sunsets, wild things were taken for granted until progress began to do away with them. Now we face the question whether a still higher "standard of living" is worth its cost in things natural, wild, and free.

In the pages that follow I will examine the wilderness system in the United States and Aldo Leopold's contribution to its creation and development. Arguably, the status of the wilderness system provides the best current evidence of how we as a nation have answered Leopold's question: whether a still higher standard of living is worth its cost in things natural, wild, and free.

Practicality versus Principle

Aldo Leopold's initial motivation toward wilderness preservation was more a matter of practicality than principle. Leopold was an

25

avid hunter. That fact is unmistakable if you read his early journals, some of which were published in the collection entitled *Round River* (*19*). "The Delta Colorado" is an example. It recounts a 1922 hunting trip to Mexico. The text is devoted almost entirely to a description of wildlife. In the first six pages there are more than 60 references to animals, birds, and fish, and fully half of them are to shooting, trapping, hunting, fishing, or eating them.

Young Aldo Leopold, sportsman and hunter, valued wilderness, not for its naturalness or for the integrity of its ecosystem, but for recreational space and for its "net production of killable bucks" (*15*). Indeed, in his early career with the Forest Service in New Mexico, Leopold was instrumental in the development and implementation of predator control policies aimed at extermination. In the same era, as editor of *The Pine Cone,* the "official bulletin of the New Mexico Game Protective Association," Leopold promoted the eradication of varmints: "Good game laws well enforced will raise enough game either for sportsmen or for varmints, but not enough for both.... The sportsmen and the stockmen," proclaimed the *Pine Cone,* "demand the eradication of lions, wolves, coyotes, and bobcats" (*7*).

Certainly, Leopold's passion for predator control would embarrass a modern ecologist or wilderness advocate, but it was hardly retrograde for its time. Leopold was pioneering the science of wildlife management, and it should surprise no one that his early efforts were sometimes crude or misguided. Indeed, even in the national parks, where hunting was prohibited, the fledgling National Park Service distinguished "good" from "bad" animals and practiced active predator control. As late as 1928, Park Service Director Horace M. Albright wrote, without apology, that "the trapping and shooting of predatory animals" adds an "element of sport into the lives of the [park] rangers" (*1*).

Game management was not Aldo Leopold's only sin against the contemporary view of wilderness as an "area where the earth and its community of life are untrammeled by man" [Wilderness Act, 1964, Sec. 2(c)]. In "The Last Stand of the Wilderness," published in 1925, Leopold argued for wilderness preservation, but not for wilderness purity or ecosystem integrity (*13*). "One wilderness area," he asserted, "could...be fitted into the National Forests of each [western] state without material sacrifice to other kinds of playgrounds or other kinds of use."

Furthermore, these wilderness areas need not be ecologically undisturbed. In 1921 Leopold had declared domestic livestock grazing an asset to wilderness "because of the interest that attaches to cattle grazing operations under frontier conditions" (*10*). Although he opposed timber harvest in wilderness when that meant building roads, he was perfectly willing to accept it in the Lake States, where the cut could be unobtrusively removed from the forest in the winter. Using this approach, he argued, "much of the remaining wild country [could be permanently used] for both wilderness recreation and timber production without large sacrifice of either use" (*13*). Fire control was also recognized as taking precedence over wilderness preservation. "Obviously," he wrote, "the construction of trails, phone lines, and towers necessary for fire control must not only be allowed but encouraged" (*13*). Indeed, even roads were approved if the control of fire required them.

In summary, the Aldo Leopold who led the fight for wilderness reservations in the national forests was not a modern ecologist seeking to preserve the integrity of natural ecosystems but a hunter and outdoorsman seeking to preserve the public hunting grounds and recreational space he cherished.

A Plea for Wilderness

To contemporary ears it may seem implausible that a man of these views should be described by James P. Gilligan (*8*) as the "father of the national forest wilderness system." Although the investigations of Donald N. Baldwin (*4*) suggest that this title should be shared with Arthur Carhart, there is no doubt that Aldo Leopold played a pivotal role in the establishment and perpetuation of wilderness reserves in the United States.

Aldo Leopold was born on January 11, 1887, in Burlington, Iowa. Leopold's family was prosperous, and both his parents were outdoor enthusiasts. Young Aldo had both the resources and the encouragement to study and appreciate nature. He pursued ornithology and natural history at the Lawrenceville School in New Jersey and later at Yale University, where he earned a graduate degree from the School of Forestry in June 1909. Shortly thereafter Leopold went to work for the Forest Service in Arizona and New Mexico.

Upon his arrival in the Southwest, Leopold observed six areas in Arizona and New Mexico that met his definition of wilderness as "a continuous stretch of country preserved in its natural state, open to lawful hunting and fishing, big enough to absorb a two weeks' pack trip, and kept devoid of roads, artificial trails, cottages, or other works of man" (*10*). The recreational potential of these six undeveloped areas was apparent to Leopold, but development threatened. By 1918 Leopold had undertaken to oppose some of the road construction that was rapidly decimating wilderness areas in the Southwest, and in 1921 he carried the wilderness campaign to his professional colleagues in the *Journal of Forestry*. In an article entitled "The Wilderness and Its Place in Forest Recreational Policy," Leopold advocated wilderness reserves in the national forests (*10*).

Public wilderness preserves were not a new idea. As early as 1832 the artist and explorer George Catlin spoke for the preservation of "Nature's works...in their pristine beauty and wildness, in a magnificent park.... A nation's Park containing man and beast" (*5*). A generation later Henry David Thoreau encouraged "national preserves ...in which the bear and panther, and even some of the hunter race, may still exist, and not be 'civilized off the face of the earth' " (*22*).

By 1921 the preservationist progeny of Catlin and Thoreau had succeeded in creating a national park system and a National Park Service to meet their needs. National forests were another matter; their lands were reserved for wise use. It was Aldo Leopold who recognized and articulated the view that wilderness preservation might be wise use. Indeed, it might be the wisest use of all for some national forest lands. He concluded that the Forest Service's historic commitment to the "doctrine of 'highest use,'" and its criterion, 'the greatest good for the greatest number'," demanded "that representative portions of some forests be preserved as wilderness" (*10*). Leopold favored wilderness areas in both forests and parks, but his argument was addressed to the Forest Service. The prototype with which he illustrated his case was the one he knew best. He called for preservation of the Gila. The *Journal of Forestry* article provided the intellectual foundation for wilderness preservation in the national forests, but it was greeted with yawns by most of his fellow foresters (*8*). Clearly, more was required.

In October of the following year Leopold formally proposed wilderness status for a large area surrounding the headwaters of the Gila River in the Gila National Forest. On June 3, 1924, District Forester Frank C. W. Pooler formally approved Leopold's wilderness proposal, reserving about 700,000 acres for wilderness as part of a district recreation plan (4, 8, 24).[1] Historians may quibble, but the Forest Service officially recognizes the designation of the Gila Wilderness on June 3, 1924, as the moment of creation for a national forest wilderness system. For the first time a federal land management agency had designated a large tract of land as wilderness recreational space, and that agency was not the Natonal Park Service.

Aldo Leopold's contribution to wilderness preservation was not limited to his role in preserving the Gila (17). Throughout the 1920s his was the nation's most active and influential voice for wilderness preservation. Leopold's 1921 article in the *Journal of Forestry* has already been mentioned (10). It was the first salvo of a modest barrage.

In 1925 alone Leopold's wilderness message reached a diverse audience. In March the popular California magazine *Sunset* published "Conserving the Covered Wagon" (12). A skit embodying the wilderness message and entitled "The Pig in the Parlor" appeared in the June 8 edition of the *Service Bulletin,* a journal for Forest Service employees. "The Last Stand of the Wilderness" appeared in the October issue of *American Forests and Forest Life* (13), and that same month the more scholarly *Journal of Land and Public Utility Economics* published "Wilderness as a Form of Land Use" (14). Finally, in November Leopold appealed to fellow hunters with his "Plea for Wilderness Hunting Grounds," published in *Outdoor Life* (11).

The words were different and skillfully tailored for each audience, but the message never varied. It was, at its core, a plea for natural and cultural diversity and an effort to stimulate effective political demand for wilderness as a critical element in that diversity.

Leopold argued that civilization is good because it creates options and opportunities, but if it conquers every acre, then options and opportunities are lost. The movement for good roads is beneficial, but "thrusting more and ever more roads into every little remaining patch of wilderness," wrote Leopold, "is sheer stupidity" (12). The pig has a multitude of beneficial uses, but we need not invite it to live in the parlor. "It is just as unwise," he wrote, "to devote 100% of the recreational resources of our public parks and forests to motorists as it would be to devote 100% of our city parks to merry-go-rounds" (14). In Leopold's view, to be truly civilized, we must pre-

---

[1]Sources vary widely with respect to acreage: more than 500,000 acres (7), 574,000 acres (21), 695,296 acres (8), more than 750,000 (24).

serve a few parks without merry-go-rounds, a few parlors without pigs, and a few wilderness areas without roads and other works of man.

Leopold's efforts may well have been necessary, but they were, in themselves, insufficient to bring about wilderness preservation on a national scale. The Forest Service was committed to scientific forestry and "wise use" of all the nation's resources, not to preservation. Indeed, at its inception under the leadership of Gifford Pinchot, the Forest Service recognized no legitimate recreational use of the forests. By 1910, however, the agency's attitude had begun to change, and by 1917 recreation was placed in the company of timber production, watershed protection, and grazing as one of the four main uses of the forests. Still, timber remained king, and a policy of wilderness preservation that threatened to remove timber from scientific management was certain to be resisted. To most foresters of his day, the attitudes and values articulated by Leopold made him different, not right.

A catalyst was required to transform Leopold's vision into Forest Service policy. That catalyst was the National Park Service. Created in 1916 and aggressively led by Steven Mather, the Park Service moved quickly to expand the park system and to monopolize the federal recreation budget.[2] The problem for the Forest Service was that most of the potential national parks were already national forests. What the Park Service would gain the Forest Service would lose. Pretty words about interagency cooperation and coordination notwithstanding, the Forest Service was not prepared to relinquish its turf without a fight.

Leaders in the Forest Service needed a strategy with which to resist national park aggrandizement, and Leopold's proposals for wilderness preservation had a certain appeal. In its zest to attract tourists and appropriations, the Park Service under Mather had resorted to quite a few cheap thrills: tunneling a park road through a giant Sequoia; feeding garbage to bears, complete with floodlights and bleachers; and pushing bonfires off a thousand-foot-high cliff, to mention just three. This kind of activity left the Park Service vulnerable to the charge that it was insufficiently committed to preserv-

---

[2]According to Arthur Carhart, Mather admitted at a 1921 national parks conference in Des Moines that he had personally intervened with the chairman of the House Appropriations Committee to prevent the allocation of funds to the Forest Service for recreational development (4).

ing nature. To be sure, the national parks had been created for the enjoyment of the American people, but they did not have to become Six Flags Over Theme World Park. Chief Forester William B. Greeley put it this way in 1927, "Let us add [an area] to the national park if that is where it belongs, but curses on the man who bisects it with roads, plants it with hotels, and sends yellow busses streaking through it with sirens shrieking like souls in torment" (9). If you had visited Yellowstone in his era, Greeley's reference would be clear.

Gradually, Greeley's Forest Service embraced the wilderness strategy. By designating wilderness areas the Forest Service was able to proclaim its commitment to recreation and preservation and to deliver the message that superlative areas of wild country were safer in the national forests than they would be if transferred to the park system. Wilderness reservations cost the Forest Service nothing in the short run. Most national forest business could continue uninterrupted, and the designations could always be rescinded later if they proved too constraining.

The wilderness strategy proved to be a success for the Forest Service and for wilderness preservation. Between 1920 and 1928 the Park Service and the Forest Service scuffled over control of approximately 2.3 million acres of national forest land. When the dust cleared, the park system had been enlarged by less than 600,000 acres, and the Forest Service had managed to retain three-fourths of what had been threatened. In the process of saving the remaining 1.7 million acres, the Forest Service created a national forest wilderness system embracing more than twice the disputed acreage (8).

By 1964, when Congress reserved to itself the power to designate wilderness areas, the wilderness strategy had been employed in many interagency conflicts. National park growth at national forest expense had been held to moderate levels, and the national forest wilderness system had grown to 14 million acres. Within the Forest Service, Aldo Leopold had articulated the wilderness message, but time and again it was fear of the Park Service that provided the incentive to act (3).

Proclamation of the Land Ethic

Aldo Leopold became a wilderness advocate because he wanted a place where he could go to play and shoot animals. Leopold worked within the Forest Service and without to promote wilderness preservation, but the same can be said for Arthur Carhart. Moreover, it was not Leopold and Carhart but fear of national park expansion that provided the primary impetus for national forest wilderness preservation. Why then should Aldo Leopold's words be canonized by the conservation community, and why should well-wishers of wilderness celebrate the centennial of his birth?

Aldo Leopold's life is worth celebrating because his wilderness achievements were real. It was Leopold's agitation that prompted preservation of the Gila. To a wilderness aficionado, that alone is cause for celebration. I made my personal celebration the evening of March 12, 1982, in Big Bear Canyon near the heart of the Gila Wilderness. A breeze whispered through the pines; a stream still fresh with snowmelt gurgled nearby. I was alone, and as I lay in my tent, I thought gratefully of Aldo Leopold and rejoiced that this place on this day had been preserved for me.

Throughout the 1920s Aldo Leopold's words both stimulated and justified wilderness preservation in the national forests. Without his wilderness vision, Forest Service strategy in meeting the national park challenge might have taken another direction entirely, and without national forest wilderness, it is doubtful that there would be a national wilderness preservation system at all.

Important as these accomplishments were, had Aldo Leopold died in 1930, I doubt that we would remember him with so much affection. It was in the last two decades of his life that Aldo Leopold expanded his ecological consciousness, proclaimed the "land ethic," and made us all his pupils.

By 1930 Aldo Leopold had been in Wisconsin six years, pursuing his work in wildlife management and no doubt reassessing his earlier efforts. In the mid-1920s, thanks in part to his successful eradication of varmints, Leopold had seen his precious Gila overrun by deer. In the absence of predators, hunting seemed the only solution. The Forest Service decided to increase hunter access by rebuilding the historic North Star Road, and the Gila Wilderness was cut in two.

The irony of this situation did not escape Leopold. Leopold the hunter still focused his attention on huntable species, but Leopold the scientist was prepared to reevaluate predation. In "Game Management in the National Forests," Leopold acknowledged that removal of large predators might be contributing to the "low net productivity of many deer herds," and concluded that "future predator control must be localized and discriminate" (*15*).

Susan L. Flader, the preeminent interpreter of Aldo Leopold's intellectual development, sees the mid-1930s as a period of disillusionment and new insight (7). Leopold was a committed forester and a pioneer in wildlife management, but a trip to Germany in the Autumn of 1935 left him shocked by the results of combining intensive forestry with intensive game management. If this was the end product of scientific management, he wanted nothing of it. Upon his return, Leopold wondered out loud if we could avoid that fate (16). "We Americans," he wrote, "have not yet experienced a bearless, wolfless, eagleless, catless woods. We yearn for more deer and more pines, and we shall probably get them. But do we realize that to get them, as the Germans have, at the expense of their wild environment and their wild enemies, is to get very little indeed?"

In September of the following year Leopold took a pack trip into real wilderness "along the Rio Gavilán in the Chihuahua sierra of northern Mexico" (7). What he observed stood in stark contrast to German forests and even to American forests north of the border. Recalling the trip a decade later, Leopold wrote, "It was here that I first clearly realized that land is an organism, that all my life I had seen only sick land, whereas here was a biota still in perfect aboriginal health. The term 'unspoiled wilderness' took on a new meaning" (18).

The contrasting experiences of 1935 and 1936 must have been humbling for Aldo Leopold, exterminator of predators, manager of forests and deer. But with the disillusionment came new insights that would ultimately be expressed in "The Land Ethic," the culmination of Leopold's growing ecological consciousness. Leopold argued for an expansion of the ethical universe to embrace "the land" and "the animals and plants that grow upon it" (20). In a fusion of science, philosophy, and aesthetics Leopold proclaimed, "A thing is right when it tends to preserve the integrity, stability, and beauty of the biotic community. It is wrong when it tends otherwise" (20). The case for wilderness preservation could hardly be put more succinctly.

"The Land Ethic" was published in 1949, the concluding essay in a collection entitled *A Sand County Almanac* (20). In the 1970s and 1980s more than a millon ecology-conscious Americans have turned to *A Sand County Almanac* for guidance, renewal, and support. This is the Leopold we know; this is the Leopold we honor. We erroneously assume the same vision that produced the "land ethic" animated the younger Leopold's wilderness crusade. It did not. Leopold's greatest contributions to wilderness preservation predate his ecological views.

Furthermore, it is doubtful that Aldo Leopold, ecologist, would have saved the Gila and encouraged the Forest Service on the path of preservation. Like John Muir a generation earlier, Leopold the ecologist might have dismissed the possibility of wilderness preserves in the ecologically and economically compromised national forests and, like Muir, concentrated his attention on the far more limited landed empire of the national parks. Indeed, it was because Leopold's crusade for wilderness preservation began as a quest for public hunting grounds that he was forced to concentrate his efforts on the national forests.

In the last analysis Aldo Leopold gave us two great gifts, and each is enhanced by the other. Aldo Leopold, forester, hunter, and outdoorsman, gave us the wilderness. Aldo Leopold, scientist and philosopher, gave us the ecological vision to understand and preserve it. That certainly is worth celebrating!

The Wilderness System

More than a century ago Henry David Thoreau described a young man's education in the forest (23). "He goes thither at first as a hunter and fisher, until at last, if he has the seeds of a better life in him, he distinguishes his proper objects, as a poet or naturalist it may be, and leaves the gun and the fish-pole behind."

Aldo Leopold made his intellectual pilgrimage. What progress have we made? At first glance the progress appears substantial.

In 1964 Congress passed the Wilderness Act, creating a national wilderness preservation system. More than nine million acres, including Leopold's precious Gila, were designated for preservation as wilderness. The federal law protecting these areas is at least as restrictive as Leopold the hunter and outdoorsman would have wanted. It prohibits timber harvest, mining, and most commercial activities. Roads, buildings, and mechanized transport are also forbidden. Primitive recreation is encouraged, and, except in the national parks, wilderness areas are open to both hunting and fishing.

Even more striking, in the 22 years since its establishment the wilderness system has grown almost tenfold. More than 400 areas are now included, and a 200,000-acre parcel near the Gila bears Aldo

Leopold's name. Today, almost 90 million acres have been set aside, and there is more to come. Within our lifetimes Americans will enjoy between 100 million and 130 million acres of designated wilderness, about five percent of the nation's land area. Eventually, we will have set aside an area equivalent to 200 Gilas.

Unfortunately, these large acreage figures obscure a less promising picture. Two-thirds of our nation's designated wilderness acreage is in Alaska; so are 18 of the 32 areas exceeding 500,000 acres — a size Aldo Leopold once described as large enough to absorb a two-week pack trip.

In 1921 young Leopold called for one large wilderness on the national forests in each of the western states (*10*). Today, only 6 of the 11 contiguous states from the Rocky Mountains west have a wilderness area of that size.

By 1925 Leopold was preaching the preservation of ecological diversity, and by the mid-1930s he was proclaiming the importance of preserving ecological integrity as well. In spite of Leopold's influence on the preservation movement, neither diversity nor integrity was specifically protected in the language of the Wilderness Act, and neither has been a dominant concern in the admission of new areas to the wilderness system. It should not be surprising, then, that by the standards of ecological diversity and ecological integrity, the national wilderness preservation system has been less than a resounding success.

Large aggregate acreage figures notwithstanding, the national wilderness preservation system fails dramatically as a representative of the nation's ecological diversity. According to George D. Davis, the United States can be divided into 233 distinct ecosystems, and only 81 of them — less than 35 percent — are adequately represented in the national wilderness preservation system (*6*).

It is equally clear that, outside of Alaska, the wilderness system has failed to protect the integrity of ecosystems. Our wilderness areas are largely rock and ice. Although they provide exceptional recreational opportunities for the photographer and backpacker, they are, for the most part, too small to provide year-round sustenance to the game species that so enthralled young Leopold and too small to preserve the predators that an older Leopold came to appreciate.

In "Escudilla" Aldo Leopold described the death of the last grizzly in New Mexico. Unless we are willing to pay the cost of things natural, wild, and free, we may live to read of the last grizzly in Montana and the last wolf in Minnesota. Then these majestic predators will be gone from the contiguous United States. Too much of the wildlife that enthused young Leopold and motivated his preservationist zeal is gone already.

In the 1940s an older and wiser Leopold understood that the preservation of a few museum pieces was not enough (20). He challenged us to a radical adjustment in our thinking. In "The Land Ethic" he wrote:

> [A] system of conservation based solely on economic self-interest is hopelessly lopsided. It tends to ignore, and thus eventually to eliminate, many elements in the land community that lack commercial value.... It assumes, falsely, I think, that the economic parts of the biotic clock will function without the uneconomic parts.

As we all know, the first rule of successful tinkering is to save *all* the parts. Wilderness may be an uneconomic part, but it is not unimportant. "[R]aw wilderness," wrote Leopold, "gives definition and meaning to the human enterprise" (20).

Half a century after Aldo Leopold, I too have hiked the Gila in New Mexico, canoed the Quetico in Ontario, and floated the Current in Missouri. As I read Leopold's accounts of these trips in *Round River* (19), I am struck by the price we have already paid for progress. Like Leopold in the 1930s, I realize that "all my life I [have] seen only sick land."

The importance of Aldo Leopold's wilderness message has not been diminished by the 40 years since his death. In one of the most moving parables from *A Sand County Almanac* Leopold challenges us to be stewards, not conquerors, of the land (20). The essay, "Thinking Like a Mountain," describes the death of a wolf:

> We reached the old wolf in time to watch a fierce green fire dying in her eyes. I realized then, and have known ever since, that there was something new to me in those eyes—something known only to her and to the mountain. I was young then, and full of trigger-itch; I thought that because fewer wolves meant more deer, that no wolves would mean a hunters' paradise. But after seeing the green fire die, I sensed that neither the wolf nor the mountain agreed with such a view.

If wilderness is to survive, this parable must speak to us all. Development is the religion of our age. Underdeveloped countries understandably strive to achieve it. Much of the western world is overdeveloped, yet we strive to extend it. Like young Leopold, we think that because less wilderness means more development, the total substitution of an artificial environment for a natural one will create a paradise on earth. I sense that neither the wolf nor the mountain agree. I hope you sense it too.

REFERENCES

1. Albright, Horace M., and Frank J. Taylor. 1928. *Oh, ranger!* Stanford University Press, Stanford, California.
2. Allin, Craig W. 1982. *The politics of wilderness preservation.* Greenwood Press, Westport, Connecticut.
3. Allin, Craig W. 1987. *Wilderness preservation as a bureaucratic tool.* In Phillip O. Foss [editor] *Federal Lands Policy.* Greenwood Press, Westport, Connecticut.
4. Baldwin, Donald M. 1972. *The quiet revolution: The grass roots of today's wilderness preservation movement.* Pruett Publishing Company, Boulder, Colorado.
5. Catlin, George. 1913. *North American Indians.* Leary, Stewart and Co., Philadelphia, Pennsylvania.
6. Davis, George D. 1984. *Natural diversity for future generations: The role of wilderness.* In James L. Cooley and June H. Cooley [editors] *Natural Diversity in Forest Ecosystems: Proceedings of the Workshop.* Institute of Ecology, University of Georgia, Athens.
7. Flader, Susan L. 1978. *Thinking like a mountain: Aldo Leopold and the evolution of an ecological attitude toward deer, wolves, and forests.* University of Nebraska Press, Lincoln. Original work published in 1974.
8. Gilligan, James P. 1954. *The development of policy and administration of Forest Service primitive and wilderness areas in the western United States.* Ph.D diss. University of Michigan, Ann Arbor.
9. Greeley, William B. 1927. *What shall we do with our mountains?* Sunset Magazine 59: 144 ff.
10. Leopold, Aldo. 1921. *The wilderness and its place in forest recreational policy.* Journal of Forestry 19: 718-721.
11. Leopold, Aldo. 1925. *A plea for wilderness hunting grounds.* Outdoor Life 56 (November): 348-350.
12. Leopold, Aldo. 1925. *Conserving the covered wagon.* Sunset 54: 21, 56.
13. Leopold, Aldo. 1925. *The last stand of the wilderness.* American Forests and Forest Life 31: 599-604.
14. Leopold, Aldo. 1925. *Wilderness as a form of land use.* Journal of Land and Public Utility Economics 1: 398-404.
15. Leopold, Aldo. 1930. *Game management in the national forests.* American Forests and Forest Life 36: 412-414.
16. Leopold, Aldo. 1936. *Naturschutz in Germany.* Birdlore 38: 102-111.
17. Leopold, Aldo. 1940. *The origin and ideals of wilderness areas.* The Living Wilderness 5: 7-9.

18. Leopold, Aldo. 1947. *Forward.* Unpublished manuscript, Aldo Leopold papers, Series 6, Box 17. University Archives, University of Wisconsin, Madison.

19. Leopold, Aldo. 1953. *Round River: From the journals of Aldo Leopold.* Oxford University Press, New York, New York.

20. Leopold, Aldo. 1970. *A Sand County almanac: With essays on conservation from Round River.* Sierra Club/Ballantine Books, New York, New York. Original work published in 1949 and 1953.

21. Nash, Roderick. 1973. *Wilderness and the American mind.* Yale University Press, New Haven, Connecticut.

22. Thoreau, Henry David. 1858. *Chesuncook.* Atlantic Monthly 2(June): 1-12; (July): 224-233; (August): 305-317.

23. Thoreau, Henry David. 1962. *Walden.* New American Library, New York, New York. Original work published in 1854.

24. U.S. Forest Service, Southwestern Region. No date. *Black Range Primitive Area with information on the Gila Wilderness and Gila Primitive Area (visitors' travel guide and map).* Albuquerque, New Mexico.

3

The Farmer as Conservationist: Leopold on Agriculture

Curt Meine

In February 1939, as part of the Wisconsin Farm and Home Week observance at the University of Wisconsin, Aldo Leopold presented an address entitled "The Farmer as a Conservationist." Leopold began his remarks with these words:

> When the land does well for its owner, and the owner does well by his land—when both end up better by reason of their partnership—then we have conservation. When one or the other grows poorer, either in substance, or in character, or in responsiveness to sun, wind, and rain, then we have something else, and it is something we do not like.
>
> Let's admit at the outset that harmony between man and land, like harmony between neighbors, is an ideal—and one we shall never attain. Only glib and ignorant men, unable to feel the mighty currents of history, unable to see the incredible complexity of agriculture itself, can promise any early attainment of that ideal. But any man who respects himself and his land can try to (5).

This quotation is vintage Leopold, displaying his characteristic mix of idealism and practicality, expressing his dual concern for the fate of man and land. It was, in fact, the first time in print that he gave his classic definition of conservation as the state of "harmony between man and land." In it too we see what Leopold had learned over a period of many years: that when one addresses the subject of agriculture, one takes on a subject of immense proportions.

For those who know of Leopold as the poet of *A Sand County Al-*

manac, or as an early voice for wilderness preservation, or as a founding father of wildlife management, it may come as a surprise to know that Leopold, while not a farmer himself, did work on a number of agricultural fronts. It is one of the less heralded aspects of his multifaceted career, but one that is bound, I believe, to become increasingly important in these times of transition on the rural landscape.

Even a cursory review of Leopold's career shows that he was involved in agriculture throughout his professional life. He spent 19 years with the U.S. Department of Agriculture as a Forest Service employee. As a pioneering formulator and practitioner of game management, he worked closely with farmers and became an expert observer of the farm landscape at a time—the 1920s and 1930s—when, like today, that landscape was undergoing great change. As a professor, Leopold taught for 15 years in one of the nation's top colleges of agriculture. As a writer, he wrote for and about farmers extensively, as much perhaps as on any topic. Finally, as a conservation philosopher, he made a special effort to define the role farming played in the greater equation of mankind's relationship to the natural environment.

The Farmer Must Do Conservation

The year was 1928. Leopold was the somewhat disgruntled associate director of the Forest Products Laboratory in Madison, Wisconsin. He had spent four years in the position, waiting for a promised promotion that never came. Through his work and writing, Leopold was already a respected figure in conservation. When word spread that he was looking for new work, opportunities quickly arose.

For more than 10 years Leopold had been devoting much of his spare time to game conservation. Game management, as we now know it, existed only in its embryonic stages. For years, Leopold had been promoting the idea, a *new* idea, that wild game could be raised on a sustained-yield basis, much as foresters raised trees. Moreover, the idea was not merely to rear game and then release it to be shot, but to manipulate habitat so that, in effect, the game raised itself. This was a radical and unproven notion, but it was an important one—and growing more important with the passing seasons. Squeezed between a vastly increased hunting public and an increasingly intensified agriculture, game populations in the 1920s were plummeting. Action had to be taken if hunting, or even casual observation of game, were to remain a viable proposition.

On May 22, 1928, Leopold signed a contract with the Sporting Arms and Ammunition Manufacturers' Institute, a consortium of major firearms manufacturers, to conduct an unprecedented survey of game conditions across the country. So primitive was the state of wildlife science that it lacked even the most basic information on game ranges, life histories, food and habitat needs, population dynamics, and susceptibility to hunting pressure. The game survey was to make at least a start in gathering such information.

The game survey constituted a major landmark in Leopold's professional development. It was an opportunity to study what had been a life-long interest—in his own words, "to make my hobby my profession." Leopold was already an astute observer of land; the game survey would hone his talent into genius. The method was straightforward: Leopold spent a month or so in each state, meeting its local experts, learning its geography, touring its backroads, talking to an amazing assortment of sportsmen, administrators, botanists, zoologists, farmers, professors, wardens, and foresters. The result, he hoped, would be a fair estimate of a state's game resources and a growing body of knowledge about game biology.

That summer of 1928, Leopold completed his first tentative surveys in Michigan, Minnesota, and Iowa. Conditions, of course, varied according to species and locality, but after his first months on the job, Leopold had begun to find evidence to support the one overriding suspicion of the times: that the sudden intensification of agriculture was eliminating the food and cover plants required by the majority of game species. Fencerows, borders, woodlots, remnant prairies, and wetlands were disappearing from the midwestern farmscape, and the quail, prairie chicken, grouse, snipe, woodcock, and in some areas even rabbits and squirrels were disappearing with them. This realization came as no surprise, but Leopold, for the first time, was giving it factual substance.

By 1929 it became apparent that the game survey as originally designed was too ambitious. Leopold and his sponsors decided to confine its coverage, at least for the time being, to the north central block of states: Minnesota, Wisconsin, Michigan, Ohio, Indiana, Illinois, Iowa, and Missouri. Leopold spent much of the next year and a half on the road, crisscrossing the Midwest, coming to know its contours with an intimacy that only grew with each new effort.

It was at this point that Leopold first began to devote his attention to the question of the farmer's role in wildlife conservation. The question had arisen before in Leopold's work and writing, but not with the same urgency. His answer was unequivocal:

> Most of what needs doing must be done by the farmer himself. There is no conceivable way by which the general public can legislate crabapples, or grape tangles, or plum thickets to grow up on these barren fencerows, roadsides, and slopes, nor will the resolutions or prayers of the city change the depth of next winter's snow nor cause cornshocks to be left in the fields to feed the birds. All the non-farming public can do is to provide information and build incentives on which farmers may act (2).

And those were the keys: to provide information and build incentives. Farmers had no more idea about the needs of game animals than anyone else, so Leopold began to write his earliest articles for farmers on the subject. The first apparently was a 1929 article, "How the Country Boy or Girl Can Grow Quail."

The second point—building incentives—proved more provocative. Throughout the 1920s, farmers were increasingly posting their lands against hunting, in order to keep their undisciplined city cousins out of their fields. The last thing farmers wanted was more game. Posting became so widespread that conscientious sportsmen were forced to come up with alternative proposals.

In 1929 and 1930, Leopold took on this issue in his work as chairman of the Game Policy Committee for the American Game Conference. The purpose of the committee, which included many of the nation's foremost game experts, was to draw up a definitive national game policy, a statement that was destined to guide the wildlife profession for the next 40 years. The policy, most of which Leopold himself wrote, was premised on the idea that "only the landowner can practice management efficiently, because he is the only person who resides on the land and has complete authority over it." A principle recommendation of the policy read:

> Recognize the landowner as the custodian of public game on all other land, protect him from the irresponsible shooter, and compensate him for putting his land in productive condition. Compensate him either publicly or privately, with either cash, service, or pro-

tection, for the use of his land and labor, on condition that he preserves the game seed and otherwise safeguards the public interest. In short, make game management a partnership enterprise to which the landowner, the sportsman, and the public each contribute appropriate services, and from which each derives appropriate support (*1*).

The important point here again is Leopold's steadfast conviction that the farmer, for reasons both practical and philosophical, was the one to *do* conservation. At the time, Leopold was speaking only of the conservation of game animals, but in the important years yet to come he would extend this notion to include nongame wildlife, plants, soil, water, and even scenic values. And it was this emphasis on individual landowner action that would lead him to be such an outspoken critic of the New Deal's top-heavy approach to conservation.

Building Communication Channels to Farmers

A great deal of Leopold's success as a conservation leader must be attributed to his unique communication skills. This was never so true as when he was working with farmers, whether in print, in the classroom, over the airwaves, or in personal contacts. This skill undoubtedly derived from his curiosity, as infectious as it was insatiable, about the land itself — its human and nonhuman denizens, its dynamic processes, its history and destiny. Many were the farmers who themselves learned to see their land more acutely as a result of Leopold's insight.

In his days as a forester, Leopold had gained a solid appreciation of rural psychology. Before getting down to business he enjoyed talking over crop prospects, soils, local lore, the vicissitudes of the weather and seasons. One day in the summer of 1931 Leopold was driving through Dane County, west of Madison, scouting potential hunting grounds for the upcoming season. Upon coming to a mail stop on the Chicago and Northwestern line known as Riley, he pulled in at a farm for a drink of water. He and the farmer, a man named Reuben Paulson, talked over their mutual concerns. Paulson needed relief from poachers and trespassers. Leopold needed a place to try out his ideas about game management. Paulson organized 11 of his neighboring farmers, while Leopold called on four of his hunting partners from town. Thus was born the Riley Game

Cooperative, an important example of the kind of cooperative management arrangement that the American Game Policy had tried to promote. Riley became a significant center for Leopold's work (as well as his recreation) for years to come, and the Riley farmers became his close friends. Leopold was, in turn, a trusted advisor to them (3).

There, of course, came a day when Aldo passed muster at Riley. He and his son Luna went to Riley one afternoon for a quail hunt and found Paulson at work in his barn, shoeing a horse. Paulson was about to drive in a nail when Aldo remarked, "Mr. Paulson, that nail is going in the wrong direction." Paulson looked at the nail, looked at Leopold, then stood and asked, "Where in the hell did you learn to shoe a horse?" He was unaware of Aldo's Forest Service credentials. "Well," Aldo replied, "I've done quite a lot of that." Paulson turned the nail around.

Riley was but one of several cooperative arrangements with farmers that Leopold helped organize and develop in the 1930s and 1940s. After Leopold joined the University of Wisconsin, these farms played an important role as study areas for the first generation of professionally trained wildlife managers. Much of the groundbreaking research in wildlife management was in fact performed on farms in southern Wisconsin. The importance of Leopold's style in these matters cannot be overstated. Even as his ecological vision sharpened in the 1930s and his conservation message became correspondingly more impassioned, Leopold never forgot that, in the midwestern farmbelt at least, it was the farmer who was on the front lines of conservation and had to be treated accordingly.

Leopold's appointment to the chair of game management at the University of Wisconsin in 1933 provided him, for the first time, a secure position from which to implement his management ideas. It is important to note that the chair was initially established within the Department of Agricultural Economics, and from this point on in Leopold's work one sees an increasing sophistication in his views on rural culture. The department was the first of its kind in the nation, and the pioneering work on rural economics, particularly that performed by his good friend George Wehrwein, would have a lasting impact on his own approaches to land use reform.

Leopold's new position entailed a number of responsibilities, including acting as a wildlife extension specialist. It was in this capacity that he served as advisor to the Coon Valley soil conservation project, the world's first watershed-wide soil erosion control effort. Leopold's interest in soil erosion dated back to the early 1920s, when as a forest inspector on the national forests in the Southwest he initiated a remarkable personal study of the ecological cause-and-effect of soil erosion on the southwestern range. With his work at Coon Valley, that interest re-emerged in the Midwest, never again to go into eclipse. Situated in the erosion-prone driftless area of western Wisconsin, Coon Valley was in 1933 a wasted watershed, ruined by deforestation, poor tillage practices, overgrazing, and soil depletion. It was, in Leopold's bitter phraseology, "one of the thousand farm communities which, through the abuse of its originally rich soil, has not only filled the national dinner pail, but has created the Mississippi flood problem, the navigation problem, the overproduction problem, and the problem of its own future continuity" (4). The work of the new Soil Erosion Service — later renamed the Soil Conservation Service — would turn the situation around through a unique program of integrated land use. In an article describing the success of Coon Valley, Leopold later wrote:

> There are two ways to apply conservation to land.
> One is to superimpose some particular practice upon the preexisting system of land use, without regard to how it fits or what it does to or for other interests involved.
> The other is to reorganize and gear up the farming, forestry, game cropping, erosion control, scenery, or whatever values may be involved so that they collectively comprise a harmonious balanced system of land use.... The crux of the land problem is to show that integrated use is possible on private farms, and that such integration is mutually advantageous to both the owner and the public (4).

Among his other extension activities as a professor of game management, Leopold instituted a short course for young farmers and presented a number of radio talks for farmers over WHA, the university-sponsored radio station. In both cases Leopold was trying to disseminate basic information on wildlife conservation to farmers. His first radio talk, delivered just after joining the university, was called "Building a Wisconsin Game Crop: Leaving Food and Cover." Others were similar: "The Farm Woodlot and the Bird Crop," "Game on the Modern Farm," and so on. The following excerpt gives the flavor of these talks:

There are many little tricks for increasing the service of woods and vegetation to wildlife. Take the grapevine, for example. A new grape-tangle on or near the ground is usually good for a new covey of quail, provided there be food nearby. How to get a new grape-tangle quickly? Select a tree with a grapevine in its top. Cut the tree but not the vine, and let it lie. In one season the vine will weave an "umbrella" over the down top which is hawk-proof and nearly manproof—a mighty fortress for bobwhite in even the deepest of snows. Leave a few cornshocks in the adjoining field and you have the "makings" of a new covey range which your friends the quail will not long overlook. One of the real mysteries of nature is the promptness with which habitable niches in the cold wall of the world are filled by living things. Our own place in the scheme of things is not the less tolerable for making room for a few of our fellow-creatures....

Your woodlot is, in fact, an historical document which faithfully records your personal philosophy. Let it tell a story of tolerance toward living things, and of skill in the greatest of all arts: how to use the earth without making it ugly.[1]

Leopold would continue to provide this sort of encouragement throughout his university years. In 1938 he began writing similar pieces for the *Wisconsin Agriculturalist and Farmer* on a variety of topics: "Plant Evergreens for Bird Shelter," "Wild Flower Corners," "Look for Bird Bands," "Windbreaks and Wildlife." A few of these, in revised form, were incorporated into *A Sand County Almanac*.

Ecology and Agriculture

When the Dust Bowl of the mid-thirties hit, Leopold was already well on his way to the fully developed ecological philosophy that would mark his mature writings in *A Sand County Almanac*. The Dust Bowl only hastened this development and led him directly to the question of what ecology had to offer by way of advice to agriculture in America. Increasingly, that advice would be framed in terms of what he called "land health": the ability of land as an integrated whole to regenerate itself. This was an issue inclusive of, but far greater than, agriculture alone.

The Dust Bowl was the upshot of the indiscriminate agricultural expansion of the post-World War I era. For Leopold and others it

---

[1]Unless otherwise noted, quotations are from materials in the University of Wisconsin Archives.

focused attention on the over-arching problem of how, in Leopold's tight phrasing, to "adjust men and machines to land." On April 15, 1935, coincidentally the day after the greatest dust storm yet swept out of the southern High Plains, Leopold delivered an address that he called "Land Pathology." In that unpublished speech he stated:

> This paper proceeds on two assumptions. The first is that there is only one soil, one flora, one fauna, and hence only one conservation problem. Each acre should produce what it is good for, and no two are alike. Hence a certain acre may serve one, or several, or all of the conservation groups. The second [assumption] is that economic and aesthetic land uses can and must be integrated, usually on the same acre. To segregate them wastes land, and is unsound social philosophy. The ultimate issue is whether good taste and technical skill can both exist in the same landowner. This is a challenge to agricultural education.

After tracing the history of destructive land use in America, Leopold in the speech asked what might be done in the social and physical sciences to hasten "the needed adjustment between society as now equipped, and land use as now practiced." The profit motive, for a number of reasons, was insufficient. Public ownership was, to a true conservative like Leopold, a last resort and impractical to boot. Legislative compulsion was unpalatable. Besides, science by this time had "shown good land use to require much positive skill as well as negative abstention." The only alternative was a kind of land ethic, and this 1935 paper was one of Leopold's important early expressions of this maturing idea. He wrote:

> I plead for positive and substantial public encouragement, economic and moral, for the landowner who conserves the public values—economic or aesthetic—of which he is the custodian. The search for practicable vehicles to carry that encouragement is a research problem, and I think a soluble one. A solution apparently calls for a synthesis of biological, legal, and economic skills, or, if you will, a social application of the physical sciences....
> I might say, defensively, that such a vehicle would not necessarily imply regimentation of private land use. The private owner would still decide what to use his land for; the public would decide merely whether the net result is good or bad for its stake in his holdings.
> Those charged with the search for such a vehicle must first seek to intellectually encompass the whole situation. It may mean something far more profound than I have foreseen.

The Dust Bowl was but one highly visible (and breathable) reminder that this sort of ethic was not merely a nice idea, but a necessary development. Leopold held that the improvement of economic tools had "exceeded the speed, or degree, within which it was good. Equipped with this excess of tools, society has developed an unstable adjustment to its environment, from which both must eventually suffer damage or even ruin. Regarding society and land collectively as an organism, that organism has suddenly developed pathological symptoms, i.e. self-accelerating rather than self-compensating departures from normal functioning. Granted that science can invent more and more tools, which might be capable of squeezing a living even out of a ruined countryside, yet who wants to be a cell in that kind of a body politic? I for one do not."

Through the latter half of the 1930s, Leopold would devote increasing amounts of his time to defining the characteristics of healthy land and tracing the implications of that definition for land use. I will refrain from discussing the ecological implications of modern agricultural systems; others have treated this subject more ably and completely than I can here. Suffice to say that, after the experiences of the 1930s, agriculture would begin to come under the scrutiny of this new science of ecology, and Leopold himself would begin to apply the precepts of ecology more stringently in his analyses. Those precepts had biological as well as social implications. On both grounds, for example, he decried in an unpublished manuscript the trend toward monotypes, warning that "the doctrine of private profit and public subsidy pushes constantly toward an extreme degree of crop-specialization, toward the grouping of uses in large solid blocks. The idea of self-sufficient farm units is submerged. The interspersion of wild and tame crops approaches zero...[producing] a landscape just as monotonous as the inherent variability of soil will permit."

By this time, of course, Leopold had himself become the owner of a worn-out farm, and he and his family had begun the process of bringing it back to life. He did not consider himself a farmer, but there is an unmistakable sense of pride in husbandry that enters his writings from this point forward, a quality evident to anyone who has read *A Sand County Almanac*.

Perhaps the finest example of this, and one most salient to this discussion, is the article to which I referred earlier, "The Farmer as a Conservationist." It is one of Leopold's forgotten masterpieces—

poignant and pointed, written in a delightful manner, and as pertinent today as it was 50 years ago.

The heart of Leopold's argument was that utilitarian motives had dominated the development of our agricultural system, to the general disadvantage of land, landowner, society, and even productivity itself. The neglect of the aesthetic qualities of land, while sounding abstract, had actually had very practical effects on the way people live on land. Again, Leopold speaks best for himself:

> If this were Germany, or Denmark, with many people and little land, it might be idle to dream about land use luxuries for every farm family that needs them. But we have excess plowland; our conviction of this is so unanimous that we spend a billion out of the public chest to retire the surplus from cultivation. In the face of such an excess, can any reasonable man claim that economics prevents us from getting a life, as well as a livelihood, from our acres?
>
> Sometimes I think that ideas, like men, can become dictators. We Americans have so far escaped regimentation by our rulers, but have we escaped regimentation by our own ideas? I doubt if there exists today a more complete regimentation of the human mind than that accomplished by our self-imposed doctrine of ruthless utilitarianism. The saving grace of democracy is that we fastened this yoke on our own necks, and we can cast it off when we want to, without severing the neck. Conservation is perhaps one of the many squirmings which foreshadow this act of self-liberation.
>
> One of the self-imposed yokes we are casting off is the false idea that farm life is dull. What is the meaning of John Steuart Curry, Grant Wood, Thomas Benton? They are showing us drama in the red barn, the stark silo, the team heaving over the hill, the country store, black against the sunset. All I am saying is that there is also drama in every bush, if you can see it. When enough men know this, we need fear no indifference to the welfare of bushes, or birds, or soil, or trees. We shall then have no need of the word conservation, for we shall have the thing itself (4).

Leopold's ideas on conservation, culture, and democracy were never so interwoven as when he addressed the topic of agriculture in American life. The Jeffersonian notion of a stable agrarian democracy of yeoman farmers had been left in the wake of the industrial revolution, but it was a buoyant ideal, and it resurfaced in Leopold's words. There was an important difference: where Jefferson had drawn his vision from his hopes for a healthy and lasting democratic republic, Leopold had had the benefit of 150 years of history and scientific advance, and his vision—deepening even as he wrote—was focused less on the polity than on the biology of healthy land. Yet, even through the intervening century and a half, the heart of the ideal remained. Freedom and individuality were still the points at issue. "The landscape of any farm," Leopold wrote, "is the farmer's portrait of himself. Conservation implies self-expression in that landscape, rather than blind compliance with economic dogma" (5).

This notion of self-expression in the farm landscape was fundamental to Leopold's thinking. He wrote in personal notes at the time, in another context:

> I expect, and hope for, a wide range of individualism as the ultimate development of the wildlife idea. There are, and should be, farmers not at all interested in shooting, but keen on forestry, or wildflowers, or birds in general. There are, and should be, farmers keen about none of these, but hipped on coons and coon dogs. I foresee, in short, a time when the wildlife on a farm will be the signature of a personality, just as the crops and stock already are. The more varied the media of individual expression, the more the collective total will add to [the] satisfaction of farm life.

That, in the end, was the focus of Leopold's work: the quality and satisfaction of farm life. To Leopold's thinking a farmscape stripped of all but its human economic components was not only at agronomic risk, but it was a waste of cultural potential. Conservation, conversely, sought to balance utility and beauty on the land; it was a challenge to use the earth without making it ugly.

A final quotation of Leopold's from a 1945 paper, "The Outlook for Farm Wildlife," speaks most directly to our farm situation today. Leopold concluded a review of trends in the farm wildlife situation by stating:

> In short, we face not only an unfavorable balance between loss and gain in habitat, but an accelerating disorganization of those unknown controls which stabilize the flora and fauna, and which, in conjunction with stable soil and a normal regimen of water, constitute land-health.
>
> Behind both of these trends in the physical status of the landscape lies an unresolved contest between two opposing philosophies of farm life. I suppose these have to be labelled for handy reference, although I distrust labels:

1. *The farm is a food-factory,* and the criterion of its success is saleable products.

2. *The farm is a place to live.* The criterion of success is a harmonious balance between plants, animals, and people; between the domestic and the wild; between utility and beauty.

Wildlife has no place in the food-factory farm, except as the accidental relic of pioneer days. The trend of the landscape is toward a monotype, in which only the least exacting wildlife species can exist.

On the other hand, wildlife is an integral part of the farm-as-a-place-to-live. While it must be subordinated to economic needs, there is a deliberate effort to keep as rich a flora and fauna as possible, because it is "nice to have around."

It was inevitable and no doubt desirable that the tremendous momentum of industrialization should have spread to farm life. It is clear to me, however, that it has overshot the mark, in the sense that it is generating new insecurities, economic and ecological, in place of those it was meant to abolish. In its extreme form, it is humanly desolate and economically unstable. These extremes will some day die of their own too-much, not because they are bad for wildlife, but because they are bad for farmers (6).

On Seeking Harmony

How do we assess Leopold's words? In the half century since he wrote, conservation has evolved into environmentalism, while farming has moved toward agribusiness. Yet one need not read far into Leopold to appreciate the timeliness—or, perhaps more accurately, the time*less*ness—of his thoughts. They remain relevant so long as people live on land and so long as the human instinct for stewardship endures. But more to the point: Do they speak to the issues of the day?

We are told today that the changes in farm tenure taking place across the continent, including and especially the foreclosure problem, represent the inexorable workings of economic trends and that the family farm has itself become an expendable commodity. We are advised to abandon the Jeffersonian view of the farmer as a sentimental holdover from a bygone era. We are asked to forget the truth, so eloquently expressed by Leopold in both word and deed, that the farm is more than a place to grow food, that farms also grow farmers, and families, and plants and animals, both wild and tame. We apply patchwork solutions to problems that have been a long time in the building and that can only be confronted by a view of history, ecology, and economics that is as wide-reaching as agriculture itself.

I think conservationists have much to offer as the nation debates these points and seeks new answers. Leopold himself was as sound a voice as one could find. He was not one to make sentimental references to "the heartland." He did not hold a romantic image of the farmer, but realized that farmers are as diverse and independently minded as any group of individuals. But he also realized that, fundamentally, a balanced society must be built on a stable system of agriculture and that this in turn must be built on an appropriate attitude toward the land that sustains us all.

His thoughts present us with a challenge. To agricultural scientists, historians, and economists, he challenges us to explore the all-too-neglected territory where separate disciplines meet.

To farmers, conservationists, and environmentalists, he challenges us to work together more than we have, to compare our aims, and to appreciate that whatever differences may exist between us pale before the common dilemma we all face as lovers of the land.

And finally, to all of us as citizens in a democracy, Leopold challenges us to consider what sort of society we wish to build: one that strives to squeeze the land for all it is worth, or one that seeks today and tomorrow the elusive harmony between humankind and land that Leopold called conservation.

REFERENCES

1. Leopold, Aldo, et al. 1931. *Report to the National Game Conference on an American Game Policy.* In *Transactions of the 17th American Game Conference.* American Game Association, Washington, D.C.
2. Leopold, Aldo. 1933. *Game cropping in southern Wisconsin.* Our Native Landscape (December).
3. Leopold, Aldo, and Reuben Paulson. 1934. *Helping ourselves: Being the adventures of a farmer and a sportsman who produced their own shooting ground.* Field and Stream 39(August): 32-33, 56.
4. Leopold, Aldo. 1935. *Coon Valley: An adventure in cooperative conservation.* American Forests 41(May): 205-208.
5. Leopold, Aldo. 1939. *The farmer as a conservationist.* American Forests 45(June): 294-299, 316, 332.
6. Leopold, Aldo. 1945. *The outlook for farm wildlife.* In *Transactions of the 10th North American Wildlife Conference.* American Wildlife Institute, Washington, D.C.

4

Aldo Leopold and the Limits of American Liberalism[1]

Roderick Nash

The Greek word *oikos,* meaning house, is the root of both economics and ecology. Over time the significance of the word shifted from the house itself to what it contained — a living community, the household. Economics, the older of the two concepts, concerns the study of how the community manages its material resources. Ecology took shape in the 1890s as the science of organisms, not just human ones, in biotic communities as they interact with each other and with their environment. Ecology, in a word, concerned communities, systems, wholes (*1, 4, 12, 47, 58*).

Because of this holistic orientation, the ecological perspective proved to be fertile soil for environmental ethics. Aldo Leopold, himself a pioneer American ecologist, made the connection explicit. "All ethics," he wrote in *A Sand County Almanac,* "rest upon a single premise: that the individual is a member of a community of interdependent parts" (*26*).

Steeped as their origins were in the social contract theory of John Locke, Americans took the concept of community to be replete with strong ethical overtones. Once you conceded that something was a member of your community, the argument for its inalienable natural rights was difficult to deny. As the definition of community

---

[1]A version of this essay appears as "Aldo Leopold's Intellectual Heritage" in J. Baird Callicott, editor, 1987, *Companion to a Sand County Almanac: Interpretive and Critical Essays,* University of Wisconsin Press, Madison. An expanded version will appear in Nash's forthcoming book *Widening the Circle: Environmental Ethics and the Limits of American Liberalism.*

broadened, so did the circle of ethical relevancy. In the century after the Revolution, American ethics underwent significant expansion. The abolition of slavery was the most dramatic example. Subsequently the rights of women, Indians, laborers, and free blacks received attention. As Bernard Bailyn recognized, the philosophy of natural rights and the attendant concept of liberty was a "contagion" so powerful in the United States that it "swept past boundaries few had set out to cross, into regions few had wished to enter" (3).

One of these unanticipated frontiers was the human relationship to that other component of the state of nature — nature itself. When ecology helped Americans think of other species and the biophysical world as an oppressed and exploited minority within the extended moral community, the contemporary environmental movement received its most characteristic insignia. Old-style conservation, plugged into American liberalism, became the new environmentalism. Environmental ethics is the latest extension of America's fundamental ideology of natural rights.

Expanding natural rights to include the rights of nature depended upon the realization that the community to which humans belonged did not end with humans. American thought anticipated this idea even before ecology provided a scientific basis for it and before Aldo Leopold pointed the way to the limits of American liberalism.

Thoreau's Holism

As a transcendentalist, Henry David Thoreau's holism stemmed from his belief in an "Oversoul" or god-like moral force that permeated everything in nature. Using intuition, rather than reason and science, humans could transcend physical appearances and perceive, as Ralph Waldo Emerson put it, "the currents of the Universal Being" binding the world together. Thoreau, in his writings (52), expressed the resulting perception: "The earth I tread on is not a dead, inert mass; it is a body, has a spirit, is organic and fluid to the influence of its spirit." In this aspect of his holism, Thoreau professed what might be termed "theological ecology" — God held things together. But the scientist in Thoreau emerged as he walked the countryside. His journals are replete with data about how organisms relate to each other and their environment, and he followed the Linnean tradition in using the phrase "economy of the universe."

Thoreau's organicism or holism, reinforced by both science and

religion, led him to remarkable ideas about his natural environment. He referred to nature and its creatures as his *society,* tran-
scending the usual human connotation of that term. "I do not," he
wrote in his journal of 1857, "consider the other animals brutes in
the common sense." He regarded sunfish, plants, skunks, and even
stars as fellows and neighbors — parts, in other words, of his community. "The woods," he declared during an 1857 camping trip in
Maine, "were not tenantless, but choke-full of honest spirits as good
as myself any day" (*51*). There was no hierarchy or any discrimina-
tion in his concept of community. "What we call wildness," he wrote
in his 1859 journal, "is a civilization other than our own." No American had thought of anything faintly resembling this before.

Although he did not use the term, an environmental ethic sprang
from Thoreau's expanded community consciousness. It began with
the axiom that "every creature is better alive than dead, men and
moose and pine trees" (*51*). He went on to question the almost om-
nipresent belief of his era in the appropriateness of human domina-
tion, kindly or not, over nature. "There is no place for man-wor-
ship," and later he wrote in his journal, "The poet says the proper
study of mankind is man. I say, study to forget all that; take wider
views of the universe."

Thoreau's expanded vision led him to rant against the Concord
farmers who engaged in the quintessentially American activity of
clearing the land of trees and underbrush: "If some are prosecuted
for abusing children, others deserve to be prosecuted for maltreat-
ing the face of nature committed to their care." Here, Thoreau
seemed to be implying that nature should have legal rights like other
minorities subject to oppression. Making the same connection on
another occasion, he pointed out the inconsistency of the president
of an anti-slavery society wearing a beaver-skin coat. While Thoreau
avoided the word "rights," his association of abused nature with
abused people placed him squarely in the path that the new environmentalists would later follow. But in the middle of the 19th century, Thoreau was not only unprecedented in these ideas, he was virtually alone in holding them.

Marsh Documents Human Impact

George Perkins Marsh, in his book *Man and Nature: Or, Physical
Geography as Modified by Human Action* (*29*), offered the first

comprehensive description in the English language of the destructive impact of human civilization on the environment. Marsh spoke 21 languages, traveled widely abroad as a U.S. diplomat, studied the history of past civilizations in detail, and wrote with passion. In the manner of the proecologists, Marsh wrote about the "balances" and "harmonies" of nature. But, unlike Thoreau, he did not move on to challenge anthropocentrism. He was perfectly content with the idea of mankind's dominion over nature, provided that it was careful and farsighted. That it had not been was the subject of his book. "Man has forgotten," Marsh thundered, "that the earth was given to him for usufruct alone, not for consumption, still less for profligate waste." Anticipating the ecological perspective of the 20th century, Marsh warned that the interrelatedness of "animal and vegetable life is too complicated a problem for human intelligence to solve, and we can never know how wide a circle of disturbance we produce in the harmonies of nature when we throw the smallest pebble into the ocean of organic life." As a corrective to previous human carelessness, Marsh proposed "geographical regeneration," a great healing of the planet beginning with the control of technology. And this, he continued, would require "great political and moral revolutions" (29).

Marsh's enormously influential book marked the first published American discussion of nature protection in ethical terms. Granted, Marsh's work contained nothing about the rights of nature; the welfare of people was uppermost in his mind. But he did suggest that human custodianship of the planet was an ethical or "moral" issue, not just an economic one. It was right, in other words, to take care of nature; wrong to exploit her.

Muir's Respect for Nature

John Muir's 1864 encounter with a Canadian flower that directed him into biocentric ethics came as the consequence of draft-dodging. President Abraham Lincoln asked for 500,000 more men to defend the Union on March 10, 1864. Muir, who was twenty-six and single, felt certain he would be called, and he apparently had no interest in fighting to save the Union or free the slaves. So Muir left his Wisconsin home in June, vanishing into the wilderness north of Lake Huron. He followed a lonely trail into a wet and darkening swamp, but suddenly there were the flowers, rare white orchids, miles from anywhere and so beautiful that he "sat down beside them and wept for joy" (*17*). Reflecting later on the experience, Muir realized his emotion sprang from the fact that the wilderness orchids did not have the slightest relevance to human beings. Were it not for Muir's chance encounter, they would have lived, bloomed, and died unseen. Nature, he generalized, must exist first and foremost for itself and for its Creator. Everything had value. "Would not the world suffer," he concluded, "by the banishment of a single weed?" (*17*).

For Muir, as for Thoreau and the earlier organicists, the basis of respect for nature was the perception of it as part of a created community to which man also belonged. God permeated Muir's environment. Not only animals but plants (like the orchids) and even rocks (Muir wrote about "crystals") and water were "sparks of the Divine Soul" (*57*). But civilization, and particularly Christianity with its dualistic separation of people and nature, obscured this truth. To reemphasize it, Muir deliberately chose organisms at the bottom of the Christian's hierarchic chain of being—like snakes. "What good are rattlesnakes for?," he asked rhetorically. And, answering his own question, replied that they were "good for themselves, and we need not begrudge them their share of life" (*35*).

Muir made a related point in reference to alligators he encountered on an 1867 hike in Florida. Though they were commonly regarded as ugly, hateful vermin, he preferred to understand the giant reptiles as "fellow mortals" filling the "place assigned them by the great Creator of us all" and "beautiful in the eyes of God." The moral implication followed: "How narrow we selfish, conceited creatures are in our sympathies! How blind to the rights of all the rest of creation!" (*36*). Here, in Muir's journal, was the first association of rights with nature in American intellectual history. Its basis lay in Muir's perception of people as members of the natural community: "Why," he asked on the hike, "should man value himself as more than a small part of the one great unit of creation?" (*33*).

Muir went on to recognize that this point of view would not be popular among his contemporaries: "A numerous class of men are painfully astonished whenever they find anything, living or dead, in all God's universe, which they cannot eat or render in some way... useful to themselves." These people, he added never think that "nature's object in making animals and plants might possibly be first of all the happiness of each one of them, not the creation of all for the happiness of one" (*36*). Muir's first published statement of this radical idea appeared in 1875. "I have never yet happened upon a trace of evidence that seemed to show that any one animal was ever made for another as much as it was made for itself" (*30, 34*). He also gave unprecedented American articulation to the organicists' vision: "When we try to pick out anything by itself, we find it hitched to everything else in the universe" (*35*).[2]

Darwin's Influence

Darwinism took the conceit out of man. In so doing, his classic works *On the Origin of Species* and particularly *The Descent of Man* constituted an important ideological source of environmental ethics. The evolutionary explanation of the proliferation of life on earth undermined dualistic philosophies at least 2,000 years old. Charles Darwin put man back into nature. He broadened the meaning of kinship. No more special creation in the image of God, no more "soul," and, it followed, no more dominion or expectation that the rest of nature existed to serve one precocious primate.

Because Darwinism built upon the holistic philosophy of the old organicists and animists, it spread rapidly through a world prepared to receive it warmly. Its influence on American ethical thinking was profound. John Muir read Darwin and could write in 1867, eight years after *On the Origin of Species,* how "this star, our own good earth, made many a successful journey around the heavens ere man was made, and whole kingdoms of creatures enjoyed existence and returned to dust ere man appeared to claim them." (*36*). Then, looking ahead, he extrapolated an idea that few among even the most enthusiastic Darwinians dared to think: "After human beings have also played their part in Creation's plan, they too may disappear without any...extraordinary commotion whatever" (*36*). For Muir evolution was an enormously humbling idea. It did, indeed, diminish human conceit, suggesting that every other creature on the planet had a right to exist — or at least to struggle to exist — equal to that of every other creature. Humans, Muir concluded, should treat nature respectfully, even reverently.

[2]The first draft of this statement, written July 27, 1869, just after Muir had his first view from the rim of Yosemite Valley, seems even more striking: "When we try to pick out anything by itself, we find that it is bound fast by a thousand invisible cords that cannot be broken to everything in the universe. I fancy I can hear a heart beating in every crystal, in every grain of sand and see a wise plan in the making and shaping and placing of everyone of them. All seems to be dancing in time to divine music." (*17*).

Darwin himself was well aware of the ethical implications of his evolutionary hypothesis. As early as 1837 he referred to animals as "our fellow brethren" and remarked that "we may be all melted together." (9). True, Darwin described fierce competition, but he saw a commonality among all the competitors. Living and dying together over the eons, everything alive participated in a universal kinship or brotherhood. It was not far from here to the idea of respect for man's fellow participants. Darwin believed that expression of this respect was one of the distinguishing marks of civilized people. In *The Descent of Man* he addressed the matter directly. In chapter 4 he discusses the idea that the moral sense, or, as Darwin preferred, "sympathies," was a product of evolution just like the eye or the hand. He believed ethics had arisen from a preethical condition where self-interest alone existed. Over time humans broadened their ethical circle to include "small tribes," then "larger communities" and eventually "nations" and "races." Eventually it reached out still further "to the imbecile, maimed, and other useless members of society." And then Darwin made an extraordinary conceptual leap: "Sympathy beyond the confines of man...to the lower animals, seems to be one of the latest moral acquisitions." Ultimately, Darwin thought that as ethics evolved all "sentient beings" would come to be included in the moral community (8).

In shaping his view on the evolution of ethics, Darwin drew, as his footnotes in *The Descent of Man* indicate, on the work of William E. H. Lecky. This distinguished Irish intellectual historian published his *History of European Models from Augustus to Charlemagne* in 1869 when Darwin was in the final stages of writing *Descent*. Lecky endeavored to show that "there is such a thing as a natural history of morals, a defined and regular order, in which our feelings are unfolded" (21). As Lecky saw it, this process had produced an improvement in moral standards since Roman times. The recent inclusion of animals in mankind's ethical circle was an important part of Lecky's evidence for this theory. Darwin notes that Lecky "seems to a certain extent to coincide" with his own conclusions (8). In fact, Lecky anticipated Darwin almost completely.

Humanitarianism

While slower-developing than its counterpart in England, the American humane movement and its components, such as vegetarian anti-vivisectionism, deserve more recognition than they have received as intellectual precursors of environmental ethics. It is impossible, of course, to deny that 19th century anti-cruelty efforts in the United States were largely confined to domesticated animals. Pain in familiar creatures was the evil to be eradicated. Moreover, humanitarians defined the problem from a human perspective. That is to say, the humane movement frequently focused not so much of the suffering of the animals as on the apparent enjoyment of that suffering by callous people. Such individuals, it pointed out, were in danger of moral atrophication. Cruelty to animals could easily extend to cruelty to other people. So the ethical principle involved became "it is wrong for humans to be cruel" rather than "cruelty violates the rights of animals." One contemporary philosopher of environmental ethics, J. Baird Callicott, goes so far as to argue that in their shallow and limited perspective the animal liberationists are not even fellow travelers with the deep ecologists and new environmentalists. He quotes an advocate of holistic environmental ethics to the effect that "the last thing we need is simply another 'liberation movement' " (6).[3]

But this contemporary criticism does historical injustice to the old humanitarians. Granted, they lacked a far-reaching ecological consciousness as well as philosophical consistency. If pain was the problem, as Mark Sagoff has argued, the animal liberationists should attempt to end it in wild as well as domestic contexts. Why not, he asks tongue in cheek, try to help the millions of wild creatures who die horrible deaths in the wilderness every day? (41). While clever, this satire is too abrupt. Clearly, the 19th century humanitarians were not biocentric; they did not differentiate between natural processes and individual organisms. But they do deserve credit for making the first stumbling steps away from an ethical system (or definition of community) that began and ended with human beings. And, logically, those first steps would be sequential — starting with the domesticated or captured laboratory creatures most familiar to humans and closest to them biologically. As Lecky and Darwin understood, there was an historical progression in the evolution of morality that began closest to home. Animals were the next in line to people. It is

---

[3]A balanced assessment of the relationship between the humane movement and environmental ethics, critical of Callicott, is the following: Warren, Mary Anne. 1983. "The Rights of the Nonhuman World." In Robert Elliot and Arran Gare, editors, *Environmental Philosophy: A Collection of Readings*, Pennsylvania State University, University Park. pp. 109-134.

not particularly helpful to fault the humanitarians, then and now, for not going further with ethical extension. What they *did do* was revolutionary enough in terms of the main currents of Western ethical thought.

Evans and Moore Anticipate Leopold

Aldo Leopold's enormous reputation often obscures the fact that he was not the first philosopher to conceive of the extension of ethics from human to environmental dimensions. The most direct American anticipations of Aldo Leopold's ethical ideas occur in the writings of Edward Payson Evans and J. Howard Moore.

Evans, a multifaceted scholar who taught modern languages at the University of Michigan from 1862 to 1867 and then moved to Germany, wrote books about animal symbolism in architecture and about the prosecution of animals in ancient and medieval courts of justice. The first product of Evans' research into ethics and psychology was an 1894 essay in *Popular Science Monthly* entitled "Ethical Relations Between Man and Beast" (*14*). It began with a statement of Evans' intent to correct the "anthropocentric assumption" in psychology and ethics, just as it had been corrected over the past several centuries in astronomy and more recently by Darwin in biology. Evans' initial task was to attack the religious basis of anthropocentrism. Anticipating Lynn White's thesis (*54*) by 70 years, Evans developed a remarkably full case against "the anthropocentric character of Christianity," which made man "a little lower than the angels" when, in reality, he is "a little higher than the ape." Evans criticized the "tyrannical mandate" contained in *Genesis* to conquer the earth, and he compares Judeo-Christianity unfavorably to the biocentric religions of the East, such as Buddhism and Brahmanism (*14*). Taken for granted by environmentalists after 1960, this exposure of the shortcomings of Western religious tradition was unanticipated in the 1890s even in the writings of so sharp a critic of orthodox Christianity as John Muir.

In the 1894 essay, but more extensively in a book published three years later entitled *Evolutional Ethics and Animal Psychology*, Evans proceeded to demonstrate the commonality of humans and animals on "strictly scientific grounds." Drawing on evolutionary biology and on the new field of animal psychology, he discussed "metempsychosis" or reincarnation, communication abilities, the aesthetic sense, and the universality of "consciousness" or what he calls "mind." In these matters Evans often skated boldly over ice that later psychologists and philosophers would find thin. But his real purpose was to exhort, and this he accomplished with unprecedented enthusiasm: "Man is as truly a part and product of Nature as any other animal, and [the] attempt to set him up on an isolated point outside of it is philosophically false and morally pernicious" (*14*). On this basis Evans branded as wrong "maliciously breaking a crystal, defacing a gem, girdling a tree, crushing a flower, painting flaming advertisements on rocks, and worrying and torturing animals." The only hint of such a far-reaching perspective in America's past was John Muir's commentary about the divine spark in plants, rocks, and water. Following the transcendentalists, he had posited a spiritual kinship between all things in the universe, but Evans stood alone in using this as the basis for "ethical relations."

Evans, who was a linguist, not a scientist, proved a better philosopher than psychologist. He carefully explained and deflated the argument that mistreatment of animals was wrong only because it tended to degrade humans. For Evans, nonhuman life forms held intrinsic rights that man must not violate. On the same grounds, Evans criticized the legal practice of regarding animal suffering as an injury done to the animal's owner. The payment of fines, he declared, did not set this wrong right.

The stated topic of his 1897 book was the evolution of ethics. Evans devoted considerable attention to the changes in morality over time. He believed he was riding the crest of an intellectual wave. "In tracing the history of the evolution of ethics," he declared, "we find the recognition of mutual rights and duties confined at first to members of the same horde or tribe, then extended to worshippers of the same gods, and gradually enlarged so as to include every civilized nation, until at length all races of men are at least theoretically conceived as being united in a common bond of brotherhood and benevolent sympathy, which is now slowly expanding so as to comprise not only the higher species of animals, but also every sensitive embodiment of organic life" (*14*).

Evans' indebtedness, even stylistically, to Charles Darwin and William Lecky is obvious. But in one respect Evans was unprecedented. Even the most vigorous of the 19th century humanitarians and natural rightists stopped with animals. Evans, however, went far beyond, apparently to every living thing and even, as observed, to such inanimate objects as rocks and minerals.

Along with Edward Evans, the turn-of-the-century American who dealt most extensively with environmental ethics was a Chicago high school teacher. J. Howard Moore came from Missouri and attended Oskaloosa College in Iowa in the 1880s. He married a sister of the renowned trial lawyer and defender of evolution, Clarence Darrow. In 1898 Moore began teaching at Crane Manual Training High School in Chicago. From this unlikely vantage point, he published a series of books on evolutionary biology and ethical philosophy that attracted international attention. Most of them are short and undocumented, apparently originating as course lectures at Crane. Moore's major statements appeared in 1906 and 1907 under the titles *The Universal Kinship* and *The New Ethics*. Henry Salt, the leading English humanitarian, published the first book, undoubtedly with the hope of closing the gap between America's nascent animal rights movement and England's well-established one. An American edition of *The Universal Kinship* appeared two years later. Salt and Moore never met but were close through correspondence until the American's death by suicide in 1916. Seven years later the Englishman dedicated *The Story of My Cousins* to Moore. The book concerned Salt's intimate relationships with cats, dogs, and birds.

Along with Darwinism, misanthropy was the driving force behind Howard Moore's ethical philosophy. A man, in his eyes, was "not a fallen god, but a promoted reptile." In fact, men were worse than snakes: "the most unchase, the most drunken, the most selfish and conceited, the most miserly, the most hypocritical, and the most bloodthirsty of terrestrial creatures" (*31*). Gradually and incompletely Moore felt, people have tried to transcend their innate selfishness, but they were still mired in what he termed the "larval stage." The extension of ethics beyond humans was a hopeful sign. Moore, in fact, understood his "New Ethics" to be the cutting edge of what he grandiloquently styled "the great task of reforming the universe" (*32*).

The starting point for Moore's ethical system, as his title implied, was universal kinship. To Moore this meant that "all the inhabitants of the planet Earth" were related "physically, mentally, morally." He took pains to explain that his sense of ethical community applied "not to creatures of your own anatomy only, but to *all* creatures" (*31*). It followed for Moore that the Golden Rule, or what he termed "The Great Law," was "a law not applicable to Aryans only, but to *all* men; and not to men only, but to *all beings.*" Moore thought of this principle as "simply the expansion of ethics to suit the biological revelations of Charles Darwin" (*32*).

At various points in his two books Moore adopted a far-reaching ethical stance. "All beings are *ends,*" he thundered. "*No* creatures are *means.*" And again: "All beings have not equal rights, but *all have rights*" (*31*). But it soon becomes clear that Moore's ethical circle had definite limits. He respected the rights of domestic animals, of course, and also of mice, turtles, insects, and fish. But plants have no status in Moore's moral system. In *The New Ethics* he explained that plants lacked "*consciousness*"—they were alive but did not "*feel.*" Consequently, they could neither be harmed nor their rights violated. Plants were "outsiders," "mere things," and not beings or creatures (*32*).

It is likely that Moore's ardent vegetarianism contributed to this attitude. But it is also pertinent that in the 1910s he would not have had the benefit of the ecologists' sense of natural process that did not attach a moral status to participation in a food chain. As it is, Moore strikes contemporary ethical philosophers as rather ridiculous, arguing, for instance, against meat eating even by wild animals! Nonetheless, much of what Moore wrote is impressive even by the standards of latter-day, ethically minded ecologists, such as Aldo Leopold, a college student when Moore's books appeared. For example, Moore wrote about "earth-life as a single process...every part related and akin to every other part." And he could state that "the *Life Process is the End—not man*" (*31*). Clearly, Moore's ideas, like those of Edward Evans, were stepping-stones toward a land or environmental ethic.

Good evolutionists as they were, both Evans and Moore saw themselves as part of an unfolding intellectual process. Their own age might be ethically unenlightened, but they could be optimistic about the future. Charles Darwin, their mentor, had been confident about the spread of altruism through natural selection. The relatively recent abolition of slavery and gains in rights for laborers and women encouraged them. Moore made the links explicit: "The same spirit of sympathy and fraternity that broke the black man's manacles and is to-day melting the white woman's chains will to-morrow emancipate the working man and the ox." Moore understood that this would not occur overnight. "New ideas," he wrote in *The New Ethics,* "make their way into the world by generations of elbowing."

But despite that lack of respect for his opinions by his contemporaries, Moore could look forward to a time when "the sentiments of these pages will not be hailed by two or three, and ridiculed or ignored by the rest; *they will represent Public Opinion and Law*" (*31*). Evans also forecast a day when "our children's children may finally learn that there are inalienable animal as well as human rights" (*14*). His use of natural rights theory and even Jeffersonian rhetoric showed the ease of transferring those concepts from humanity to nature. And his sense of timing was uncanny. The "children's children" of Evans' generation lived after 1960, and some of them celebrated Aldo Leopold and became deep ecologists and new environmentalists.

Ecology Widens the Circle

One of the problems with the "theological ecology" of a John Muir or a Henry David Thoreau was its basis on faith rather than fact. God provided the ultimate glue that bound everything to everything. Darwin, of course, offered abundant scientific reason for believing all life to be interrelated in its origins, but he did not push on to explore the ways in which that relationship was effected. He, too, let the vague presence of a Creator do what science could not yet accomplish.

Beginning in the 1890s, however, ecologists such as Henry C. Cowles, Frederic Clements, and Charles Elton went into the field, looked at nature as a whole, and tried to explain its interrelationships and interdependencies. They wrote about biological "communities," "webs" of life, and "ecosystems." A few of them moved on to recognize the ethical implications. Cornell University Horticulturist Liberty Hyde Bailey concluded three decades of what he termed a "biocentric" approach to his subject with the publication in 1915 of *The Holy Earth* (*2*). The book followed a well-worn intellectual path in arguing that abuse of the earth was morally wrong because it was God's creation. Bailey went on to advocate abandoning "cosmic selfishness" and developing a sense of "earth righteousness." This, he felt, would help human beings "put our dominion into the realm of morals. It is now in the realm of trade." At the University of Chicago, Alfred Emerson thought that in its study of how individual interests gave way to those of the group ecology had the potential to provide "a scientific basis for ethics" (*13*).

Interdependence, as a description of nature and as a basis for determining human conduct toward nature, rose with the rise of ecology. Walter P. Taylor, president of the Ecological Society of America, commented that "there is little rugged individualism in nature" and went on to depict the ecosystem as a "closely organized cooperative commonwealth of plants and animals" (49). In 1936 Secretary of Agriculture Henry Wallace remarked that his generation needed a "Declaration of Interdependence" just as the colonists had required a Declaration of Independence" (50). Wallace's observation was perceptive. In the 1770s democratic-republican theory provided the rationale for extending the notion of rights, at least in theory, to all men. In the early 20th century ecology suggested reasons for broadening the concept of community and the consequent notions of rights and of ethical behavior still further.

Schweitzer's Reverence for Life

Probably because he couched his philosophy in simple terms and lived it dramatically in the heart of Africa, Albert Schweitzer had a far-reaching impact on the development of environmental ethics in the United States. Born in the Alsace-Lorraine region between France and Germany in 1875, Schweitzer, at age 30, abruptly resigned his university appointments in philosophy and theology and his career as a concert organist to become a doctor of medicine in Africa.

In September 1915, while on a steamer moving up the Ogowe River, Schweitzer discovered in the phrase "Reverence for Life" the most valid basis for ethics. Schweitzer built a theory of value based on the "will-to-live" that he understood every living being to possess. Right conduct for a human consisted of giving "to every will-to-live the same reverence for life that he gives to his own" (44).

Schweitzer made it abundantly clear that his reverence for life did not end with human beings. In fact, he said, "the great fault of all ethics hitherto has been that they believed themselves to have to deal only with the relations of man to man." In his eyes, "a man is ethical only when life, as such, is sacred to him, that of plants and animals as that of his fellow men" (44). Elsewhere Schweitzer went still further, apparently extending his ethics to all matter. The ethical person, Schweitzer wrote in 1923, "shatters no ice crystal that sparkles in the sun, tears no leaf from its tree, breaks off no flower, and is careful not to crush any insect as he walks" (*46*). So Schweitzer would place a worm, washed onto pavement by a rainstorm, back into the grass and remove an insect struggling in a pool of water. In a 1935 essay he called for "making kindness to animals an ethical demand, on exactly the same footing as kindness to human beings." But this demanded "so great a revolution for ethics" that philosophers had hitherto resisted making the conceptual leap. This was particularly the case in Europe and North America where ethics traditionally concerned relationships between people. Schweitzer made it his life work to inspire the thinking out of the details of "the ethic of love for all creation" (*46*). One result was his celebrated acts of compassion toward animals and insects. In this way he felt he partially discharged the moral obligation incurred by virtue of his good fortune compared to other beings. The powerful and privileged status humans enjoyed in the natural community entailed for Schweitzer not a right to exploit but a responsibility to protect.

Like William Lecky and Charles Darwin, Schweitzer concerned himself with the history and future of ethics. He believed in the potential of ethical evolution. He wrote that the thoughtful person must "widen the circle from the narrowest limits of the family first to include the clan, then the tribe, then the nation, and finally all mankind" (*45*). But this was only the beginning for Schweitzer. "By reason of the quite universal idea...of participation in a common nature, [one] is compelled to declare the unity of mankind with all created beings." Of course, Schweitzer understood that so fundamental an intellectual revolution as the extension of ethics to new categories of beings was not easy. And World War I reminded him of the shortfall in even person-to-person ethics. But he took hope from the history of ideas: "It was once considered stupid to think that colored men were really human and must be treated humanely. This stupidity has become a truth" (*45*). In the same manner Schweitzer predicted in 1923 that the circle would continue to widen: "Today it is thought an exaggeration to state that a reasonable ethic demands constant consideration for all living things down to the lowliest manifestations of life. The time is coming, however, when people will be amazed that it took so long for mankind to recognize that thoughtless injury to life was incompatible with ethics" (*43*).

Albert Schweitzer's ideas reached the United States in English translations of his books in the 1920s and 1930s. Although his was a mystical holism, it coincided remarkably with the ecologists' concept of a biotic community. No life was worthless or merely instrumental to another life. Every being had a place in the ecosystem and, some philosophers and scientists were beginning to think, a right to that place.

Leopold's Contribution

Few today would challenge Aldo Leopold's reputation as one of the seminal thinkers in the modern American statement of environmental ethics. Yet the gist of what he called "the land ethic" and the basis of his enormous reputation amounts to but 25 small and undocumented pages at the conclusion of a book he did not live to see in print—*A Sand County Almanac* (*26*). Nevertheless, within two decades Leopold's statement became the intellectual touchstone for the most far-reaching environmental movement in American history. In 1963 Secretary of the Interior Stewart L. Udall declared that "if asked to select a single volume which contains a noble elegy for the American earth and a plea for a new land ethic, most of us at Interior would vote for Aldo Leopold's *A Sand County Almanac*" (*53*). Callicott called Leopold "the father or founding genius of recent environmental ethics," a writer who created the standard or "paradigm" of an ethical system that included nature. Wallace Stegner thought of *A Sand County Almanac* as "one of the prophetic books, the utterance of an American Isaiah" (*48*), and Donald Fleming, Harvard's historian of ideas, called him "the Moses of the New Conservation impulse of the 1960s and 1970s, who handed down the Tablets of the Law but did not live to enter the promised land" (*16*). In a similar vein, another tribute singles out Leopold as "an authentic patron saint of the modern environmental movement, and *A Sand County Almanac* is one of its new testament gospels" (*42*). Van Rensselaer Potter dedicated a 1971 volume to Leopold as one "who anticipated the extension of ethics to Bioethics" (*38*). *A Sand County Almanac*, which Leopold thought might never even find a publisher, sold a million copies in several paperback editions after his death.

Born in comfortable circumstances in Burlington, Iowa, in 1887, Aldo Leopold's early experience with hunting and ornithology inclined him toward an outdoor-oriented profession. He pursued it at

Yale, graduating from that university's school of forestry in 1909. It was an exciting time to launch this kind of career. President Theodore Roosevelt and his Chief Forester Gifford Pinchot had succeeded in making a new idea called "conservation" a keystone of progressive politics. Their well-publicized governors' conference on the conservation of natural resources, held at the White House, took place just as Leopold graduated from college. Indeed, the graduate program he entered in the fall of 1908 owed its existence to the generosity of Gifford Pinchot. Understandably, Leopold absorbed much of the utilitarianism of Pinchot and the pioneer conservationists. Nature was to be used — albeit wisely and efficiently — for the greatest good of the greatest number (of people, of course) in the longest possible run.

Management was the mecca of utilitarian conservation, and Leopold started his professional life in 1909 as a manager of national forests in Arizona and New Mexico. One of his first projects was a campaign to exterminate predators, chiefly wolves and mountain lions, in the interest, he then believed, of helping the "good" animals, cattle and deer. But the emergence of the ecological sciences brought a new perspective. As he matured, Leopold absorbed its import. He came to believe that "the complexity of the land organism" was "the outstanding scientific discovery of the twentieth century," and he realized that predators were part of the whole. By 1933, when he assumed a professorship of wildlife management at the University of Wisconsin, Leopold could tell his students that the entire idea of good and bad species was the product of narrow-minded human bias. One of his lecture notes stated that "when we attempt to say that an animal is 'useful,' 'ugly' or 'cruel,' we are failing to see it as part of the land. We do not make the same error of calling a carburetor 'greedy.' We see it as part of a functioning motor."[4] On another occasion he advised those who would modify the natural world that "to keep every cog and wheel is the first precaution of intelligent tinkering" (*28*).

In this organic conception of nature, species functioned like organs within a body, or, following one of Leopold's favorite metaphors, like parts of an engine. It was one of the hallmarks of 20th century ecology and a foundation of environmental ethics.

[4]Leopold, Aldo. "Wherefore Wildlife Ecology?" Undated lecture notes, Aldo Leopold papers, Box 8, University of Wisconsin Archives, Madison, Wisconsin.

Groping for another way to make the point, Leopold chose as the title of a 1944 essay the arresting phrase "thinking like a mountain." It described an afternoon, years before, when Leopold and his Forest Service crew were lunching on a cliff overlooking a New Mexican river. Suddenly they saw a wolf cross the current, and, operating under the old ethical criteria, instantly opened fire. "I was young then," Leopold remembered, "and full of trigger-itch; I thought that because fewer wolves meant more deer, that no wolves would mean hunters' paradise." The wolf fell, but Leopold scrambled off the rimrock in time to "watch a fierce green fire dying in her eyes" (26). The green fire haunted his thought for 30 years. It started him toward the realization that wolves and other predators were necessary for the healthy herds of game animals that humans prized. But, beyond that, perhaps the wolf was a legitimate part of the southwestern ecosystem. Perhaps its presence had ecological and ethical, if not immediate economic, justification. Perhaps Leopold and his trigger-happy partners had not taken what Henry David Thoreau called "the wider view." For Leopold this transcendence of anthropocentrism was thinking like a mountain.

Aldo Leopold's first exploration of the ethics of the human relationship to nature appeared in an unpublished paper written in 1923, when he was an assistant director of the national forests in Arizona and New Mexico. Entitled "Some Fundamentals of Conservation in the Southwest," the essay began traditionally enough with the assumption of the need for the "development" of the region and the importance of "economic resources" in that process (27).[5] For most of his paper Leopold drew on the familiar Pinchot-inspired position that conservation was necessary for continued prosperity. But in a remarkable conclusion he turned to "conservation as a moral issue." To be sure, Pinchot, Roosevelt, and especially W. J. McGee had used similar rhetoric, but only in the sense of the morality of equal human rights to resources. This amounted to the familiar democratic rationale for Progressive conservation.

Leopold, however, had something else in mind. The argument that the earth was man's "physical provider" and, hence, worthy of ethical consideration left him unsatisfied. He wondered if there was not a "closer and deeper relation" to nature based on the idea that

[5]The essay was finally published in 1979, as cited, thanks to the efforts of Susan L. Flader and Eugene Hargrove.

the earth was *alive*. With this concept Leopold moved into unchart-
ed waters. The animal rightists or humanitarians, both in England
and the United States, were clearly concerned with living things.
But what about geographical features, such as oceans, forests, and
mountains? Were they animate or inanimate, living or merely mechanical? Intuitively, Leopold rebelled against the idea of a "dead
earth." He already knew enough about ecology to understand the
importance of interconnections and interdependencies. Somehow
this rendered hollow the traditional distinction between organic and
inorganic things.

In his search for help with these concepts Leopold found, rather
surprisingly, the Russian philosopher Peter D. Ouspensky. Almost
an exact contemporary of Leopold, Ouspensky published *Tertium
Organum* in 1912. An English translation appeared in the United
States in 1920 (*37*), and Leopold quotes from it in his 1923 essay.
The quotations are accurate, but Leopold, characteristically, provided no title or page references and three times misspelled the
author's name as "Onspensky." Nonetheless, what excited Leopold
about Ouspensky was the Russian's conviction that "there can be
nothing dead or mechanical in Nature...life and feeling...must exist
in everything." So, the philosopher continued, "*a mountain, a tree,
a river, the fish in the river, drops of water, rain, a plant, fire* — each
separately must possess a mind of its own." Ouspensky actually
wrote about "the mind of a *mountain*," and it seems certain that
Leopold remembered this phrase 20 years later when he titled his
own essay concerning the wolf.

Ouspensky based his views on the assumption that everything in
the universe had a "phenomena," or visible appearance, and a
"noumena." The latter was hidden to humans, and Ouspensky
variously described it as life, emotions, feeling, or mind. Leopold,
although a scientist, had sufficient confidence in his intuition to
grasp this idea and went on to accept Ouspensky's argument that
combinations of objects and processes could also be said to have a
life of their own. The whole was greater than the sum of the parts.
So cells functioned together to make organs and arrangements of
organs made possible organisms. But Ouspensky did not stop here.
Many organisms, working together in the context of air, water, and
soil, constituted a superorganism with its own particular noumena.
Such functioning communities could not be divided without destroying their collective lives, or, as Ouspensky put it in a phrase

Leopold quoted, "anything indivisible is a living being." Take away the heart, for example, and you kill the greater life of the wolf. Remove the wolf from the ecosystem and you alter the noumena of the biotic community of which it was a part. The erosion of soil produces the same alteration. The conclusion Ouspensky drew and Leopold applauded was that the earth itself—or, as Leopold came to prefer, land—was not dead but alive. With his superior writing skills, Leopold came to the Russian's aid in expressing the concept: the earth was alive, "vastly less alive than ourselves in degree, but vastly greater than ourselves in time and space—a being that was old when the morning stars sang together, and, when the last of us has been gathered unto his fathers, will still be young" (27).

For Leopold in 1923 the Ouspensky-supported assumption that the earth was, in Leopold's words, "an organism possessing a certain kind and degree of life" offered reason enough for an ethical relationship. "A moral being," he simplified the matter, "respects a living thing." This proposition, of course, could and would receive intensive scrutiny by later philosophers. But Leopold, already probing the scientific basis of a functioning earth-organism, did not pursue the philosophical puzzles. As he saw it, the "indivisibility of the earth—its soil, mountains, rivers, forests, climate, plants, and animals" was sufficient reason for respecting the earth "not only as a useful servant but as a living being" (27).

Ten years passed before Aldo Leopold wrote again about the ethical dimension of conservation. When he did so, Susan Flader thinks, he wrote "in a strikingly different manner"—as an ecologist rather than a metaphysician and theologian (15). Her assessment is true in part. As Leopold turned in the early 1930s from a government to an academic career and associated with renowned ecologists like Charles Elton, he absorbed a new vocabulary of chains, flows, niches, and pyramids. The glue holding the earth together consisted of food and energy circuits rather than divine forces or noumena. But there are striking continuities extending from the 1923 essay. The seeds of the key concepts in Leopold's land ethic are all present in the early paper. He had discovered the idea that a life community extended far beyond traditional definitions. He had argued for an ethical relationship to both its component parts and to the whole. And he had found that a strictly economic posture toward nature created serious ecological and ethical problems. Leopold's plunge into ecology represented not so much a switch as an extension. He never stopped working on the borderline between science and philosophy, using each to reinforce the other. When science lost sight of the broad picture in a welter of detail, philosophy adjusted the focus. Perhaps Leopold remembered Ouspensky's warning about the tendency of scientists to "always study the little finger of nature" (*37*). Ecologists, at any rate, were the most likely scientists to meet holistic-thinking philosophers half way.

The next building block in Aldo Leopold's land ethic was a paper read in New Mexico on May 1, 1933. Published as "The Conservation Ethic" (*23*), its major contribution was the idea of ethical evolution. Like so many commentators on this subject, Leopold noted the parallels between human slavery and unconditional ownership of land. The fact that slavery had been challenged and abolished encouraged him with regard to nature. "The Conservation Ethic," then, begins with a reference to "god-like Odysseus" who returned to his Greek homeland to hang, on a single rope, a dozen slave-girls accused of misbehavior during his absence. Yet, Leopold explained, Odysseus was an ethical man, and he did not condone murder. The point was that slaves were property and, as such, outside Odysseus' ethical community. Relations with them were strictly utilitarian, "a matter of expediency, not of right and wrong." With the passage of time, Leopold continued, an "extension of ethics" occurred. Slaves became people, not property; the abolition of slavery followed. But, and this was Leopold's thesis, "there is as yet no ethic dealing with man's relationship to land and to the non-human animals and plants which grow upon it. Land, like Odysseus' slave-girls, is still property." The progress of civilization still entails "the enslavement of...earth." Leopold hoped that the conservation movement represented an awareness that "the destruction of land...is wrong." And, Leopold made clear, he did not mean "wrong" in the sense of inexpedient or economically disadvantageous. He meant it in the same sense that abuse of another human being was wrong (*23*).

Nowhere in the 1933 essay did Leopold refer to the rights of land or nonhuman life. Ethics were the ideas or ideals of people that acted as restraints on people. Leopold defined an ethic as a "limitation on freedom of action in the struggle for existence." In other words ethics applied to situations in which a person who could have done a particular action held back because he knew that action was wrong. Sometimes, Leopold understood, this involved working directly against immediate self-interest. The ethical person foregoes the opportunity to improve his economic position by robbing another person. In the same way Leopold hoped that a land ethic might be a constraint against robbing or exploiting the land. Although subsequent philosophers have worried the point to near death, Leopold simply dismissed the notion that animals, plants, and soil had ethical relations *with* people. For Leopold it was a one-way street: human beings were the ones who did the restraining, who extended *their* ethics to include nature (*23*). Several authors recently have dissected Leopold's philosophy on these points (*5, 22, 58*).

Aldo Leopold gave no specific indication in his 1933 statement (or in *A Sand County Almanac*) that anyone had ever thought about expanding ethical sequences before his time. Yet he must have known that a fellow biologist, Charles Darwin, had in 1871 written extensively on the subject. In fact, in identifying "the tendency of interdependent individuals or societies to evolve modes of cooperation" known as ethics, Leopold nearly plagiarized Darwin. He also ignored his intellectual debt to William Lecky, the historian Darwin credited with calling initial attention to ethical evolution as well as to Edward Evans, J. Howard Moore, Alfred North Whitehead, and Albert Schweitzer. Yet the work of these men was readily available when Leopold was in college or early in his professional career. As an ecologist Leopold took his ethics further than most of them — to collections of organisms and habitats organized as ecosystems — but it is disconcerting that this scientist, so meticulous in his recording of biological facts, would play so loosely with historical ones. Similarly surprising is the occasional tendency of Leopold scholars to aggrandize their subject at the expense of historical accuracy. Callicott, for example, overstates the case with his opinion that Leopold's ideas are "the first self-conscious, sustained and systematic attempt in modern Western literature to develop an ethical theory which would include non-human natural entities and nature itself in the purview of morals" (*7*).

In late 1947 and early 1948 Leopold reviewed his 1923 and 1933 essays, added insights from subsequent papers of 1939 and 1947, and wrote for *A Sand County Almanac* a final chapter entitled "The Land Ethic." It begins with the story of Odysseus and the slave-girls, lifted with only minor changes from the 1933 paper, and the concept of ethical evolution. But then Leopold turns with fresh insights to the origin and meaning of ethics. Ethics, he explains, derive from the recognition that "the individual is a member of a community of interdependent parts." On the one hand the individual competes within this community, but "his ethics prompt him also to cooperate (perhaps in order that there may be a place to compete for)." The land ethic, then, "changes the role of *Homo sapiens* from conquerer of the land-community to plain member and citizen of it. It implies respect for his fellow-members, and also respect for the community as such." Behind this sentence lies Leopold's recognition that while, in one sense, humans are simply members of a "biotic team," in another their technologically-magnified power to impact nature sets them apart from the other members. So human civilization needed the restraints afforded by a land ethic. Just as concepts of right and wrong had made human society more just, Leopold felt they would enhance justice between species and between man and the earth. The entire import of *A Sand County Almanac*, Leopold writes in his foreword, is directed to helping land "survive the impact of mechanized man." His statement of the basic problem and solution is characteristically pithy and powerful: "We abuse land because we regard it as a commodity belonging to us. When we see land as a community to which we belong, we may begin to use it with love and respect" (26).[6]

Frequently in *A Sand County Almanac*, as well as in his other writings, Leopold took an instrumental view of the land ethic. It embodied restraints that help humans live a happy and healthy life. It is prudent to be ethical with regard to the natural order that sustains the human one. A battle-scarred veteran of conservation policy wars, Leopold knew this was the best way to sell his philosophy in the 1930s and 1940s. He also knew it was not the full story. His most radical ideas, and his greatest significance for the 1960s and beyond, concern the *intrinsic* rights of nonhuman life forms and of life communities or ecosystems to exist. Early in "The Land Ethic" Leopold affirms "the right to continued existence" of not only animals and plants but waters and soils as well. And he writes that the life forms that share the planet with people should be allowed to

---

[6]Two later essays by Leopold that expanded on "The Land Ethic" chapter in *A Sand County Almanac* can be found in references *24, 25*. The best analyses of *A Sand County Almanac* are those of Peter A. Fritzell (*18*) and the contributors to *Companion to A Sand County Almanac*, edited by J. Baird Callicott and to be published in 1987 by the University of Wisconsin Press.

live "as a matter of biotic right, regardless of the presence or absence of economic advantage to use." This means "there are obligations to land over and above those dictated by self-interest," obligations grounded on the recognition that humans and the other components of nature are ecological equals (26).

This idea of "biotic right" was the intellectual dynamite in *A Sand County Almanac*. The Darwinian evolutionists and the old-style humanitarians had occasionally glimpsed the idea of a morality that extended beyond human society, but Leopold, with the aid of ecology, gave it its most dramatic articulation at least to the late 1940s. Most of the earlier advocates of extended ethics dealt almost exclusively with individual organisms and then, generally, with the higher animals. This is not to denigrate them. Theirs was an expected intellectual way-station on the road away from anthropocentric ethics. Leopold's achievement was to follow the road to its termination in ecosystems, environment or, as he preferred, "land." He is correctly regarded as the most important source of *environmental* ethics.

Implications of a Land Ethic

If Darwin killed dualism, the ecologists presided over its burial. Humans were simply one of many members of a greatly expanded biotic community. The moral implications of this idea for human behavior were, to say the least, problematic. Philosophers after Leopold would devote hundreds of pages to the subject. But Leopold was quite clear about what he thought the land ethic mandated in terms of behavior. It did not, at the outset, mean having *no* impact on one's environment. As a biologist Leopold knew this was an impossibility for any organism. He would have chuckled at vegetarians, such as Henry Salt, and extreme right-to-life sects, such as the pre-Christian Indian group called the Jains who, following the philosophy of *Ahimsa*, breathed through gauze so as not to inhale living microorganisms. Even Albert Schweitzer's assistance of struggling worms and insects would have struck Leopold as naive and beside the main point of land health. Leopold's concept of Schweitzer's reverence-for-life principle was precisely that—for life *in toto* and not so much for the individual players in life processes.

Leopold would have approved of Schweitzer's principle of taking life only for essential purposes and, then, with reverence for that which was killed. According to Leopold, hunting, meat eating, even, in Leopold's words, "the alteration, management and use" of the ecosystem were acceptable. The essential proviso, Leopold wrote as early as 1933, was that any human action be undertaken in such a way as "to prevent the deterioration of the environment." By 1948, when he finished "The Land Ethic" for *A Sand County Almanac,* Leopold had refined this principle into what has become his most widely-quoted axiom. A land use decision "is right when it tends to preserve the integrity, stability, and beauty of the biotic community. It is wrong when it tends otherwise" (*26*).[7]

Leopold was well aware of the massive obstacles standing in the way of ethically directed human relations to the environment based on an ethic that extended to the limits of the ecosystem. "No important change in ethics," he wrote in *A Sand County Almanac,* "was ever accomplished without an internal change in our intellectual emphasis, loyalties, affections, and convictions." The conservation movement of his day, the 1940s, had not, in his view, touched these "foundations of conduct." As proof, Leopold submitted that "philosophy and religion have not yet heard" of "the extension of social conscience from people to land." In this belief Leopold was both correct and in error. Although evidently unrecognized by Leopold, philosophy, biology, history, religion, and even law (the humane legislation) had all heard of the extension of ethics, if not as far as Leopold's proposal at least beyond human-to-human interactions. But Leopold was correct in assuming that Western thought in general contained little approaching the holistic character of his moral philosophy. Still, in matters of slowly shifting attitudes and values, he knew the value of existential patience. Ethics, after all, were ideals, not descriptions of how people actually behaved. "We shall never achieve harmony with land," Leopold believed, "any more than we shall achieve justice or liberty for people. In these higher aspirations the important thing is not to achieve, but to strive" (*28*).

This parallel that Leopold drew between human-to-human and human-to-nature ethics found its way into many of Leopold's final essays. In "The Ecological Conscience," he wrote, it "has required 19 centuries to define decent man-to-man conduct and the process is

[7]Leopold's first published version of this axiom appeared in 1947 in "The Ecological Conscience" (*25*), and it stressed the common membership of humans and other life forms in one community. The 1947 version reads: "A thing is right only when it tends to preserve the integrity, stability, and beauty of the community, and the community includes the soil, waters, fauna, and flora, as well as people." Extensive analysis of this idea appears in several other published works (*19, 39, 40*).

only half done; it may take as long to evolve a code of decency for man-to-land conduct." His prescription for such progress was not to allow economics to dictate ethics: "Cease being intimidated by the argument that a right action is impossible because it does not yield maximum profits, or that a wrong action is to be condoned because it pays." That philosophy, Leopold concluded, "is dead in human relations, and its funeral in land-relations is overdue" (25).[8]

Aldo Leopold's pessimism concerning public comprehension, not to speak of acceptance, of the land ethic was supported by the early history of *A Sand County Almanac*. As an unpublished typescript, it was sent to and rejected by so many publishing houses that the author despaired of ever seeing his work in print. Leopold did not live to see the reviews of *Sand County*, but they probably would have disappointed him. Most critics understood the book to be just another collection of charming nature essays. Very few reviewers even recognized the ideas that a later generation would find compelling. Initial sales of the slender green volume were slow before its renaissance in paperback in the late 1960s.

Environmental Ethics after Leopold

The most obvious reason for the initial lack of public interest in Leopold's ideas was their truly radical quality. What he proposed would have necessitated a complete restructuring of basic American priorities and behavior. Also involved was a radical redefinition of progress. The conquest and exploitation of the environment that had powered America's westward march for three centuries was to be replaced as an ideal by collaboration and coexistence. The land ethic, in short, placed unprecedented restraints on a process that had won the West and lifted the nation to at least temporary greatness as a world power. Taken literally, Leopold's philosophy ended with abruptness the accustomed freedom with which Americans had hitherto dealt with nature. It was no longer a one-way street; the land had rights too.

The American mood of the late 1940s and early 1950s was not likely to receive these proposals with enthusiasm. The deprivations of the Great Depression fed immediately into those of World War

---

[8]Underscoring the importance of ecology to his ethical system, Leopold used the term "ecological conscience" here to describe what in *A Sand County Almanac* he called "the land ethic."

II. Emerging from 15 years of frustrated materialism, Americans pursued the main chance with extraordinary vigor. The postwar decade was a time for building homes and families. Upholding the integrity, stability, and beauty of the ecosystem and granting even nonutilitarian species biotic rights made little sense to the first wave of baby boomers.

But most ecologists also rejected Leopold's ideals. As Donald Worster has shown, ecology after 1950 became increasingly abstract, quantitative, and reductionist (58). Crop yields, not the organic wholeness of life and matter, set the new tone. Interdependence gave way as an organizing concept to efficiency, just as in the heyday of Gifford Pinchot's Progressive conservation. Meanwhile, a substantial proportion of life scientists turned their attention inward, to cells and molecules. The kind of integrative biology or natural history at which Aldo Leopold excelled seemed hopelessly old-fashioned.

In this relatively hostile climate of opinion the ethical consequences of an ecological perspective took what shelter they could find in the humanities and the softer sciences. Joseph Wood Krutch, a literary critic turned desert naturalist, called in 1954 for a "large morality" predicated on the realization that people "must be part not only of the human community but of the whole community,... the natural...community." Citing Leopold as well as Albert Schweitzer, Krutch made clear his belief in the "right to live" of creatures that had no direct bearing on human welfare (20).

Another ecologist who retained the interests of a natural historian, Rachel Carson, published the best-selling *Silent Spring* in 1962. The book warned the nation about the dangers of indiscriminate use of pesticides to kill insects. Its emphasis was clearly anthropocentric, but Carson's closest associates felt that Schweitzer's principle of reverence for life motivated all her writing. Significantly, she dedicated *Silent Spring* to Albert Schweitzer. More than any other writer, Carson acquainted Americans of the 1960s with the fundamentals of the ecological perspective and its ethical implications.

By the 1970s and 1980s, a few professional ecologists were returning to an interest in moral philosophy. Edward O. Wilson, the Harvard entomologist, believed he understood the process. As he saw it, the first investigators of a complex subject, such as ecology, raised ethical questions. But the increase of knowledge in the field produced a focus on facts that Wilson called "amoral." But "finally, as understanding becomes sufficiently complete, the questions turn ethical again." Wilson hoped the environmental sciences were, at the conclusion of the 20th century, ready to proceed to this third phase. Indeed, he believed that "the future of the conservation movement depends on such an advance in moral reasoning" (55).

Hoping to make his own contribution to this development, Wilson focused on modern technological man's ability to reduce biological diversity by the wholesale extermination of species. "This," Wilson felt, "is the folly our descendants are least likely to forgive us." He explained his opinion first on the instrumental level. It was wrong to eliminate species because mankind might find them useful in the future—for foods, medicines, and the like. Wilson regarded this as a "surface ethic," but, like Leopold, he was fully prepared to use it to advance the conservation cause. "The only way to make a conservation ethic work," he believed "is to ground it in ultimately selfish reasoning...people will conserve land and species fiercely if they foresee a material gain for themselves, their kin and their tribe" (56). At the surface level, then, ethics arose from the old Darwinian necessity for survival.

But Wilson, like Krutch, did not personally believe this rationale for conservation was enough. In 1984 he wrote about a "deep conservation ethic" based on "biophilia," which Wilson defined as the tendency of the human mind to "affiliate" with other forms of life and with the life process. He was speaking here about psychological, not physical, survival. He granted that the subject was still cloaked in mystery. The core of the issue seemed to be that the human mind had evolved in association with myriad life forms and, even if subconsciously, needs them for the continued survival of "the human spirit." It was a question of kinship, Wilson explained, and respect for the 10 billion "bits" of genetic information in even the humblest living creature. Borrowing Schweitzer's term, Wilson concluded that "reverence for life" would one day be understood in terms of evolutionary biology and evolutionary psychology. His definition of community, then, not only ranged across the entire contemporary ecosystem but backward in time to the beginning of evolution. Man belonged, physically and psychologically, to both past and present ecosystems. Wilson's achievement was to use this expanded definition of community as a reason for respecting the rights of the other members.

The impact of ecology on modern ethical philosophy may, finally, be illustrated by the work of David Ehrenfeld. A professor of ecology at Rutgers University and associate editor of *Human Ecology*, Ehrenfeld's career illustrates Wilson's point about the tendency of a scientific discipline to shift over time toward a philosophical orientation. Ehrenfeld's doctoral research concerned a narrow biochemical problem. But in the late 1970s he began to explore the ethical implications of ecological science. The problem, he felt, could be distilled into a single phrase: "the arrogance of humanism." Ehrenfeld used "humanism" here to denote a bias in favor of the human species and against other species. It had the same connotation as racism or sexism, and the antidote in each case was to widen the ethical community (*10, 11*).

Now Ehrenfeld, like Wilson, was prepared to concede that protection of the human interest was a useful rationale for conservation. "Selfishness, within bounds," he believed, "is necessary for the survival of any species, ourselves included." But Ehrenfeld perceived technologically enhanced anthropocentrism to have "grown ugly and dangerous. Humanism...must now be protected against its own excesses." This is the point at which ethics became relevant as a restraining device. Climaxing several centuries of Anglo-American thought on the matter, Ehrenfeld declared that "long-standing existence in Nature" carries with it "the unimpeachable right to continued existence." He stressed that this ethical precept was not in the least predicated on the usefulness of a species to humans. This, in fact, was where old-style "resource" conservation with its economically oriented arguments had failed. What Ehrenfeld wanted was recognition of the rights of "nonresources," including species that had no significance to people or even to the healthy functioning of the ecosystem. Leopold, as Ehrenfeld pointed out, had made much of the healthy functioning argument, but he saw himself as going further. He wanted a land ethic that would include even those members of the biotic community whose disappearance could not possibly affect land health. As an example, Ehrenfeld put forward the furbish lousewort, "a small member of the snapdragon family which has probably never been other than a rare constituent of the forests of Maine."[9] In its case Ehrenfeld thought the only firm base for ethical respect was "existence value," a kind of ecological be-

[9]At the time of Ehrenfeld's publications the furbish lousewort had received extensive publicity in connection with a Maine hydroelectric project that would have flooded its only known habitat.

cause-it-is-there attitude (*11*).

To make this point most dramatically Ehrenfeld cited a 1976 article by physician Bernard Dixon on smallpox. He pointed out that the virus, once the scourge of mankind, had been so effectively pursued by world health organizations that its only existence was as a guarded laboratory specimen. Vaccination programs had eliminated the germ in its only natural habitat: human beings. Dixon pointed out that "this is the first time in history when man has been able to obliterate—for all time and by conscious rational choice—a particular form of life," and he wondered if there was "a case for the conservationists to move in and call a halt?" Both Dixon and Ehrenfeld believed that such a case could be made but for different reasons. The physician pointed to the research value of smallpox and its potential use in fighting other human diseases. The ecologist, however, cited smallpox as part of the biotic community, a product of evolution just like wolves and whales and redwood trees. There was no logical reason, and here the men agreed, for discriminating against a form of life just because it was small and harmful to humans. Such reasoning drew the fullest implications of ideas René Dubos had introduced 20 years earlier. It was ethically appropriate, from Ehrenfeld's perspective, to protect the right to life of even an organism whose only function was to prey on people. Here was surely the ultimate submergence of the human ego in the ecological community.

Aldo Leopold felt that acceptance of a land ethic depended upon changing long-established cultural attitudes. In 1948 he was not optimistic about the prospects for such a change. But in the next two decades a dramatic rise in public understanding of ecological realities created fertile soil for the growth of environmental ethics. Ecology substituted a new biological basis of community for the old theological organicism. It suggested many practical reasons for enlarging the moral community to include nature, but also created an argument for the intrinsic rights of other species and even of the environment as a whole. Undoubtedy, Leopold would have been both surprised and pleased at the appearance of "bioethics" as a scholarly field, at the rise of environmental philosophy, and the appearance, in 1979, of an entire journal, *Environmental Ethics*, devoted to further exploration of many of the ideas he raised 40 years before. He would have applauded the several endangered species acts of the 1970s, which declared that certain nonhuman members of an extended American community had, quite literally, a right to life, liberty, and the pursuit of happiness. Leopold certainly would have been sympathetic with the "deep ecology" movement and, while it was not his style, perhaps encouraged some aspects of radical environmentalism as interpreted by action-oriented groups, such as Earth First! and Greenpeace. He would have read with satisfaction Christopher Stone's 1974 essay *Should Trees Have Standing? Toward Legal Rights for Natural Objects,* which singled him out as the source of the no-longer-unthinkable idea that the land deserved recognition in courts of law just as do human litigants. While Leopold died discouraged about ethical extension, the next generation of ecologists felt that "the idea of rights conferred by other-than-human existence is becoming increasingly popular" (*10*). Indeed, by the centennial of his birth, 1987, Aldo Leopold would have noticed that, thanks to the emergence of an ecological perspective and of environmental ethics, increasing numbers of Americans were extending the limits of their nation's traditional liberalism to include nature.

REFERENCES

1. Allee, W. C., Alfred Emerson, Thomas Park, Orlando Park, and Karl Schmidt. 1949. *Principles of animal ecology.* W. B. Saunders Company, Philadelphia, Pennsylvania.
2. Bailey, Liberty Hyde. 1915. *The holy earth.* Macmillan, New York, New York.
3. Bailyn, Bernard. 1967. *The ideological origins of the American Revolution.* Harvard University Press, Cambridge, Massachusetts.
4. Brewer, Richard C. 1960. *A brief history of ecology.* Occasional paper 1. C.C. Adams Center for Ecological Studies, Kalamazoo, Michigan.
5. Callicott, J. Baird. 1979. *Elements of an environmental ethic: Moral considerability and the biotic community.* Environmental Ethics 1(Spring): 71-81.
6. Callicott, J. Baird. 1980. *Animal liberation: A triangular affair.* Environmental Ethics 2(Winter): 311-338.
7. Callicott, J. Baird. 1983. *The land aesthetic.* Environmental Review 8(Winter): 345-358.
8. Darwin, Charles. 1874. *The descent of man, and selection in relation to sex.* Murray, London, England.
9. Darwin, Francis, editor. 1887. *The life and letters of Charles Darwin.* D. Appleton and Company, New York, New York.
10. Ehrenfeld, David. 1976. *The conservation of non-resources.* American Scientist 64(November-December): 648-656.
11. Ehrenfeld, David. 1978. *The arrogance of humanism.* Oxford University Press, New York, New York.
12. Elton, Charles S. 1966. *The pattern of animal communities.* Methuen & Company, Ltd., London, England.
13. Emerson, Alfred. 1946. *The biological basis of social cooperation.* Transac-

tions, Illinois Academy of Science 39(May): 9-18.

14. Evans, Edward Payson. 1897. *Evolution ethics and animal psychology.* Appleton, New York, New York.

15. Flader, Susan L. 1979. *Leopold's some fundamentals of conservation: A commentary.* Environmental Ethics 1(Summer): 143-148.

16. Fleming, Donald 1972. *Roots of the new conservation movement.* Perspectives in American History 6: 7-91.

17. Fox, Stephen. 1981. *John Muir and his legacy: The American conservation movement.* Little, Brown & Co., Inc., Boston, Massachusetts.

18. Fritzell, Peter A. 1976. *Aldo Leopold's a Sand County almanac and the conflicts of ecological conscience.* Transactions, Wisconsin Academy of Sciences, Arts and Letters 64(Fall): 22-46.

19. Heffernan, James D. 1982. *The land ethic: A critical appraisal.* Environmental Ethics 4(Fall): 235-247.

20. Krutch, Joseph Wood. 1954. *Conservation is not enough.* American Scholar 23(Summer): 295-305.

21. Lecky, William E. H. 1955. *History of European morals from Augustus to Charlemagne.* George Braziller, Inc., London, England.

22. Lehmann, Scott. 1981. *Do wildernesses have rights?* Environmental Ethics 3(Summer): 129-146.

23. Leopold, Aldo. 1933. *The conservation ethic.* Journal of Forestry 31(October): 634-643.

24. Leopold, Aldo. 1939. *A biotic view of land.* Journal of Forestry 37(September): 727-730.

25. Leopold, Aldo. 1947. *The ecological conscience.* Bulletin of the Garden Club of America (September): 45-53.

26. Leopold, Aldo. 1949. *A Sand County almanac and sketches here and there.* Oxford University Press, New York, New York.

27. Leopold, Aldo. 1979. *Some fundamentals of conservation in the Southwest.* Environmental Ethics 1(Summer): 131-141.

28. Leopold, Luna, editor. 1953. *Round River: From the journals of Aldo Leopold.* Oxford University Press, New York, New York.

29. Marsh, George Perkins. 1965. *Man and nature: Or, physical geography as modified by human action.* Harvard University Press, Cambridge, Massachusetts.

30. Mighetto, Lisa. 1985. *Muir among the animals.* Sierra 70(March-April): 69-71.

31. Moore, J. Howard. 1906. *The universal kinship.* Bell and Sons, London, England.

32. Moore, J. Howard. 1909. *The new ethics.* Block, Chicago, Illinois.

33. Muir, John. 1875. *Wild wool.* Overland Monthly 14(April): 361-366.

34. Muir, John. 1901. *Our national parks.* Houghton Mifflin, Boston, Massachusetts.

35. Muir, John. 1911. *My first summer in the Sierra.* Houghton Mifflin, Boston, Massachusetts.

36. Muir, John. 1916. *A thousand-mile walk to the Gulf.* Houghton Mifflin, Boston, Massachusetts.

37. Ouspensky, P. D. 1981. *Tertium organum: The third canon of thought, a key to the enigmas of the world.* (Translated by E. Kadloubovsky). Routledge and Kegan Paul, Ltd., London, England.

38. Potter, Van Rensselaer. 1971. *Bioethics: Bridge to the future.* Prentice-Hall, Englewood Cliffs, New Jersey.

39. Regan, Tom. 1984. *The case for animal rights.* University of California Press,

Berkeley, California.

40. Regan, Tom, editor. 1984. *Earthbound: New introductory essays in environmental ethics.* Temple University Press, Philadelphia, Pennsylvania.

41. Sagoff, Mark. 1984. *Animal liberation and environmental ethics: Bad marriage, quick divorce.* Osgood Hall Law Journal 22(Summer): 297-307.

42. Schoenfeld, Clay. 1978. *Aldo Leopold remembered.* Audubon 80(May): 28-37.

43. Schweitzer, Albert. 1950. *Animal world: Jungle insights into reverence for life.* Beacon Press, Boston, Massachusetts.

44. Schweitzer, Albert. 1933. *Out of my life and thought: An autobiography.* Oxford University Press, New York, New York.

45. Schweitzer, Albert. 1936. *Indian thought and its development.* (Translated by Mrs. C.E.B. Russell.) Henry Holt and Company, New York, New York.

46. Schweitzer, Albert. 1947. *Philosophy of civilization: Civilization and ethics.* (Translated by John Naish). A&C Black, Ltd., London, England.

47. Siry, Joseph V. 1984. *Marshes of the ocean shore: Development of an ecological ethic.* Texas A&M University Press, College Station, Texas.

48. Stegner, Wallace. 1985. *Living on our principal.* Wilderness 48(Spring): 15-21.

49. Taylor, Walter P. 1935. *Significance of the biotic community in ecological studies.* Quarterly Review of Biology 10(September): 291-307.

50. Taylor, Walter P. 1936. *What is ecology and what good is it?* Ecology 17(July): 333-346.

51. Thoreau, Henry David. 1868. *The Maine woods.* Ticknor and Fields, Boston, Massachusetts.

52. Thoreau, Henry David. 1906. *Writings of Thoreau* (20 volumes). Houghton Mifflin, Boston, Massachusetts.

53. Udall, Stewart L. 1963. *The quiet crisis.* Holt, Reinhart & Winston, Inc., New York, New York.

54. White, Lynn. 1967. *The historical roots of our ecological crisis.* Science 155(March 10): 1,203-1,207.

55. Wilson, Edward O. 1984. *Biophilia.* Harvard University Press, Cambridge, Massachusetts.

56. Wilson, Edward O. 1984. *Million-year histories: Species diversity as an ethical goal.* Wilderness 48(Summer): 12-17.

57. Wolfe, Linnie Marsh, editor. 1938. *John of the mountains: The unpublished journals of John Muir.* Houghton Mifflin Company, Boston, Massachusetts.

58. Worster, Donald. 1977. *Nature's economy: The roots of ecology.* Sierra Club, San Francisco, California.

Adding brick to the chimney at the Shack, 1938

5

The Scientific Substance
of the Land Ethic[1]

J. Baird Callicott

Aldo Leopold was not the *only* giant in the republic of American conservation letters to suggest that nature be included within the purview of ethics. Roderick Nash is correct to point out this fact in his contributions to this volume and to *Companion to A Sand County Almanac* (*16*). Neither was Leopold the *first* to propose, in Nash's terms, that "nature has rights too." Henry David Thoreau advocated something like that. So did John Muir and the German humanitarian Albert Schweitzer.

Aldo Leopold's *unique* contribution to this celestial chorus of voices crying in and for the wilderness was to provide a sound *scientific* foundation for a land or environmental ethic.

Thoreau's environmental ethic ultimately rested on the mystical Transcendental or Emersonian belief that a divine presence permeated all natural objects (*15*). Muir often used the same Transcendental rhetoric on behalf of nature's moral value. In Muir's florid prose, sequoia groves were "cathedrals," the Hetch Hetchy Valley a "temple" consecrated for worship, trees were "psalm-singing," and mountains extended "glad tidings" (*13*). If anything, Muir outdid Emerson and certainly the more geogenous Thoreau in Transcendental enthusiasm for human communion and moral integration with nature.

---

[1]This chapter contains excerpts from the essay "The Conceptual Foundations of the Land Ethic" in J. Baird Callicott, editor, *Companion to A Sand County Almanac: Interpretive and Critical Essays*, to be published in 1987 by the University of Wisconsin Press, Madison.

For all the spiritual appeal of his borrowed Transcendental ideas and the contagious enthusiasm of his Transcendental rhetoric, Muir developed a more compelling argument for an environmental ethic that was all his own invention. It was based on conventional Judeo-Christian doctrine: God created all forms of life, and in His view they are all good right down to the "smallest transmicroscopic creature that dwells beyond our conceitful eyes and knowledge" (14). God bestows upon each and every creature, Muir went on to argue, "the same species of tenderness and love as is bestowed on angels in heaven or saints on earth" (14). He must care, Muir pointed out, for the creation as a whole, not just for one precocious part: "Why should man value himself as more than a small part of the one great unit of creation?," Muir demanded to know, "[a]nd what creature of all that the Lord has taken pains to make is not essential to the completeness of that unit — the cosmos?" (14).

As a philosopher who has struggled for more than a decade to formulate a coherent and effective environmental ethic, I can testify that Muir's theocentric environmental ethic is especially elegant and powerful. Intrinsic value, the most direct and demanding basis of respect and moral consideration, is provided for nature as a whole and nonhuman natural entities simply by divine fiat. Exploring the same complex of ideas further (although to my knowledge Muir himself never publicly did so), in establishing His creation God may be understood to have made *species* and to care for each as such, while *specimens* are understood to come and go. Hence, we are not compelled to extend "rights" to individual plants and animals, nor are we prohibited from the usufruct of nature, from harvesting the bounty of nature for human benefit. Rather, we must take care to preserve "the" alligator and "the" rattlesnake — two species favored by Muir for illustrative purposes because they were so often vilified, "according to closet researchers of clergy" — and all the other exuberant kinds of life that together compose that one great integrated and dynamic unit: the creation (14).

Albert Schweitzer, prominent among the host of lesser figures whom Nash mentions as having anticipated Leopold in advocating the extension of ethics beyond the confines of human-to-human relationships, turned to the Orient for a philosophical foundation for his environmental ethic. In classical Hindu metaphysics the self-same essence — *Atman* or *Brahman* — lies at the kernel of all phenomenal things (5). In the Hindu-inspired philosophy of Arthur

Schopenhauer, clearly the immediate source of Schweitzer's founda-
tional notions, this hidden essential identity underlying all things
was called the "will-to-live" (20). Schweitzer more clearly (but less
subtly) than Schopenhauer developed a neoKantian Voluntarist
version of *ahimsa*—the Hindu doctrine of compassion and non-
injury with respect to all living things. According to Schweitzer (21):

> Just as in my own will-to-live there is a yearning for more life, and
> for that mysterious exaltation of the will-to-live which is called plea-
> sure, and terror in the face of annihilation and that injury to the will-
> to-live which is called pain; so the same obtains in all the will-to-live
> around me, equally whether it can express itself to my comprehen-
> sion or whether it remains unvoiced.
>
> Ethics thus consist in this, that I experience the necessity of prac-
> ticing the same reverence for life toward all will-to-live, as toward my
> own. Therein I have already the needed fundamental principle of
> morality. It is *good* to maintain and cherish life; it is *evil* to destroy
> and check life.

The Right Stuff

Granted the appeal that these several mystical, theological, and
metaphysical philosophies may have for those whom William James
called the tender-minded, they don't play in Peoria—or in the
White House, or on Capitol Hill, or on Wall Street, or in the other
centers of American political and economic power. A David Brower,
or Huey Johnson, or Gaylord Nelson in the thick of a real-world
legal and political battle over the preservation of free-flowing rivers,
wilderness, or endangered species would invite derision and certain
defeat were he to rest his case on Emerson's Transparent Eyeball, or
God's charge to Adam to dress the garden and keep it, or on Indra's
jeweled net and the reverence for life it implies. But such stalwart
and politically savvy environmental advocates as these have all in-
voked—and with considerable effect—Aldo Leopold's land ethic.
Why?

Because *A Sand County Almanac* happened to be handy when
the quiet crisis came to public attention? Was Leopold simply lucky
to have been, posthumously, in the right place at the right time?
Thoreau was more widely read and studied then (as now). Muir
wrote in a more impassioned style—rhetorically more in tune with
the general excess of the late 1960s and early 1970s. And the invoca-

tion of Schweitzer could have brought to bear the venerable dignity
of the continental intelligentsia and the tremendous moral authority
of his medical ministry to darkest Africa. Was Leopold just acciden-
tally or arbitrarily selected to be the moral standard bearer of the
contemporary environmental movement?

No, *A Sand County Almanac* has become the bible of modern
conservation for good reason—because science talks. And it talks,
more especially, the language of the tough-minded who invariably
sit in the seats of political, economic, and judicial power. Aldo Leopold's great philosophical and political achievement was not *that* he
made a case for an extension of ethics from human-to-human relationships to the humanity-to-nature relationship. Rather, it was *how*
he made that case. He expressed the land ethic beautifully and
elegantly to be sure, but he also expressed it exclusively in the
vocabulary of science. Hence, the land ethic—unlike the
Transcendental nature ethic of Thoreau or the theocentric creation
ethic of Muir, or the reverence-for-life ethic of Schweitzer, or, for
that matter, the several ecologically blind versions of animal welfare
ethics that Nash also throws in as antecedents of the land ethic—
cannot be dismissed with a wink and a chuckle as romantic non-
sense, religious naivete, mystical poppycock, or uninformed senti-
mentalism.

From its scientific conceptual foundations the land ethic gains its
objectivity, universality, authority, and power. From its author it
gains its credibility. Aldo Leopold was a lifetime member of the conservation establishment. He was a passionate sportsman; he was an
employee of the U.S. Forest Service (that most conservative and en-
vironmentally unregenerate of conservation agencies, stamped in
perpetuity with the utilitarian "resourcism" of Gifford Pinchot); and
he was the founder of the modern theory of "game" management.
He knew whereof he spoke. He spoke with the voice of a professional. And he spoke the argot of 20th century management science,
which he imaginatively turned to the service of a brand new nature
ethic.

The Ethical Sequence

"The Land Ethic" opens with a charming and poetic evocation of
Homer's Greece, the point of which is to suggest that today land is
just as routinely and remorsely enslaved as people were then. A pan-

oramic glance backward to our most distant cultural origins, Leopold suggests, reveals a slow but steady moral development over
three millennia. More of our relationships and activities ("fields of
conduct") have fallen under the aegis of moral principles ("ethical
criteria") as civilization has grown and matured. If moral growth
and development continue, as not only a retrospective review of ancient history but recent past experience would suggest, future generations will censure today's casual and universal environmental bondage as today we censure the casual and universal human bondage of
3,000 years ago.

A cynical critic might scoff at Leopold's sanguine portrayal of
human history. Slavery survived as an institution in the "civilized"
West, more particularly in the morally self-congratulatory United
States, until a mere generation before Leopold's own birth. And
Western history from imperial Athens and Rome to the Spanish In-
quisition and Third Reich has been a disgraceful series of wars, per-
secutions, tyrannies, pogroms, and other atrocities.

The history of moral practice, however, is not identical with the
history of moral consciousness. Morality is not descriptive; it is
prescriptive or normative. In light of this distinction, it is clear that
today, despite rising rates of violent crime in the United States and
institutional abuses of human rights in Iran, Chile, Ethiopia, Guate-
mala, South Africa, and many other places, and despite persistent
organized social injustice and oppression in still others, moral consciousness is expanding more rapidly now than ever. Civil rights,
human rights, women's liberation, children's liberation, fetus liberation, animal liberation, etc., all indicate, as expressions of newly
emergent moral ideals, that ethical consciousness (as distinct from
practice) has, if anything, accelerated of late, thus confirming
Leopold's historical observation.

Leopold next points out that "this extension of ethics, so far
studied only by philosophers" — and, therefore, the implication is
clear, not very satisfactorily studied — "is actually a process in ecological evolution" (11). What Leopold is saying here simply is that
we may understand the history of ethics, fancifully alluded to by
means of the Odysseus vignette, in biological as well as philosophical
terms. From a biological point of view, an ethic is "a limitation on
freedom of action in the struggle for existence" (11).

The phrase "struggle for existence" unmistakably calls to mind
Darwinian evolution as the conceptual context in which a biological

account of the origin and development of ethics must ultimately be located. And at once it points up a paradox: Given the unremitting competitive "struggle for existence," how could "limitations on freedom of action" ever have been conserved and spread through a population of *Homo sapiens* or their evolutionary progenitors?

For a biological account of ethics, as Harvard social entomologist Edward O. Wilson recently wrote, "the central theoretical problem ...[is] how altruism [elaborately articulated as morality or ethics in the human species] which by definition reduces personal fitness, could possibly evolve by natural selection" (*9, 26*). According to modern sociobiology, the answer lies in kinship. But according to Darwin, who tackled this problem himself "exclusively from the side of natural history" in *The Descent of Man,* the answer lies in society (*4*). And it was Darwin's classical account (and its divers variations), from the side of natural history, that informed Leopold's thinking in the late 1940s.

The Origin of Ethics

Let me put the problem in perspective. How, we are asking, did ethics originate and, once in existence, grow in scope and complexity?

The oldest answer in living human memory is theological. God (or the gods) imposes morality on people. And God (or the gods) sanctions it. A most vivid and graphic example of this kind of account occurs in the Bible when Moses goes up on Mount Sinai to receive the Ten Commandments directly from God. That text also clearly illustrates the divine sanctions (plagues, pestilences, droughts, military defeats, etc.) for moral disobedience. On-going revelation of the divine will, of course, as handily and simply explains subsequent moral growth and development. (John Muir, as I just mentioned, forcefully pointed out that among the divinely appointed duties that we may find in the Bible is our obligation to respect and care for our fellow creatures and for the creation as a whole.)

Secular Western philosophy, on the other hand, is almost unanimous in the opinion that the origin of ethics in human experience has somehow to do with human reason. Reason figures centrally and pivotally in the "social contract theory" of the origin and nature of morals in all its ancient, modern, and contemporary expressions, from Protagoras, to Hobbes, to Rawls. Reason is the wellspring of virtue, according to both Plato and Aristotle, and of categorical imperatives, according to Kant. In short, the weight of Western philosophy inclines to the view that we are moral beings because we are rational beings. The ongoing sophistication of reason and the progressive illumination it sheds upon the good and the right explains "the ethical sequence," the historical growth and development of morality, noticed by Leopold.

A scientist, however, cannot be satisfied with either of these general accounts of the origin and development of ethics. The idea that God gave morals to man is ruled out in principle, as any supernatural explanation of a natural phenomenon is ruled out in principle in natural science. And while morality might *in principle* be a function of human reason (as, say, mathematical calculation clearly is), to suppose that *in fact* it is would put the cart before the horse. Reason appears to be a delicate, variable, and recently emerged faculty. It cannot, under any circumstances, be supposed to have evolved in the absence of complex linguistic capabilities that depend, in turn, for their evolution upon a highly developed social matrix. But we cannot have become social beings unless we assumed limitations on freedom of action in the struggle for existence. Hence, I suggest we must have become ethical before we became rational.

Extending Parental and Filial Affections

Darwin, probably because of reflections somewhat like these, turned to a minority tradition of modern philosophy for a moral psychology consistent with and useful to a general evolutionary account of ethical phenomena. A century earlier Scottish philosophers David Hume and Adam Smith had argued that ethics rest upon feelings or "sentiments," which, to be sure, may be both amplified and informed by reason (*10, 24*). Because feelings or sentiments are arguably far more common or widespread than reason in the animal kingdom, they would be a far more likely starting point for an evolutionary account of the origin and growth of ethics.

Darwin's account, to which Leopold unmistakably (if elliptically) alludes in "The Land Ethic," begins with the parental and filial affections common, perhaps, to all mammals (*4*). Bonds of affection and sympathy between parents and offspring permitted the formation of small, closely kin social groups, Darwin argued. Should the parental and filial affections bonding family members chance to extend to less closely related individuals, that would permit an enlargement of the family group. Should the newly extended community more successfully defend itself and/or more efficiently provide for itself, the inclusive fitness of its members would be increased, Darwin reasoned. Thus, the more diffuse familial affections that Darwin (echoing Hume and Smith) calls the "social sentiments" would be spread through a population (4).

Human morality, properly speaking—morality as opposed to mere altruistic instinct—requires, in Darwin's terms, "intellectual powers" sufficient to recall the past and imagine the future, "the power of language" sufficient to express "common opinion," and "habituation" to patterns of behavior deemed, by common opinion, to be socially acceptable and beneficial (4). Even so, ethics proper, in Darwin's account, remains firmly rooted in moral feelings or social sentiments that—no less than physical faculties, he expressly avers—were naturally selected, by the advantages for survival and especially for successful reproduction, afforded by society (4).

The proto-sociobiological perspective on ethical phenomena, to which Leopold as a natural historian was heir, leads him to a generalization that is remarkably explicit in his condensed and often merely resonant rendering of Darwin's more deliberate and extended paradigm: Since "the thing [ethics] has its origin in the tendency of interdependent individuals or groups to evolve modes of cooperation, ...[a]ll ethics so far evolved rest upon a single premise: that the individual is a member of a community of interdependent parts" (11).

Hence, we may expect to find that the scope and specific content of ethics will reflect both the perceived boundaries and actual structure or organization of a cooperative community or society. *Ethics and society or community are correlative.* This single, simple principle constitutes a powerful tool for the analysis of moral natural history, for the anticipation of future moral development (including, ultimately, the land ethic) and for systematically deriving the specific precepts, the prescriptions and proscriptions, of an emergent and culturally unprecedented ethic like a land or environmental ethic.

The Ethics-Community Connection

Anthropological studies of ethics reveal that the boundaries of the moral community are generally coextensive with the perceived boundaries of society (22). The peculiar (and from the urbane point of view sometimes inverted) representation of virtue and vice in tribal society—the virtue, for example, of sharing to the point of personal destitution and vice of privacy and private property—reflect and foster the life way of tribal peoples (19). Darwin, in his leisurely and anecdotal discussion, paints a vivid picture of the intensity, peculiarity, and sharp circumscription of "savage" mores: "A savage will risk his life to save that of a member of the same community, but will be wholly indifferent about a stranger" (4). As Darwin portrays them, tribespeople are at once paragons of virtue "within the limits of the same tribe" and enthusiastic thieves, man-slaughterers, and torturers without (4).

For purposes of more effective defense against common enemies, or because of increased population density, or in response to innovations in subsistence methods and technologies, or for some mix of these or other forces, human societies have grown in extent or scope and changed in form or structure. Nations, like the Iroquois nation or Sioux nation, came into being upon the merger of previously separate and mutually hostile tribes. Animals and plants were domesticated and erstwhile hunter-gatherers became herders and farmers. Permanent habitations were established. Trade, craft, and later industry flourished. With each change in society came corresponding and correlative changes in ethics. The moral community expanded to become coextensive with the newly drawn boundaries of societies; and the representation of virtue and vice, right and wrong, good and evil changed to accommodate, foster, and preserve the economic and institutional organization of new social orders.

Today we are witnessing the painful birth of a human supercommunity, global in scope. Modern transportation and communication technologies, international economic interdependencies, international economic entities, and nuclear arms have brought into being a "global village." It has not yet become fully formed and it is at tension—a very dangerous tension—with its predecessor, the nation-state. Its eventual institutional structure, a global federalism or whatever it may turn out to be, is, at this point, completely unpredictable. Interestingly, however, a corresponding global human ethic, the "human rights" ethic as it is popularly called, has been more definitely articulated.

Most educated people today pay lip-service at least to the ethical precept that all members of the human species, regardless of race, creed, or national origin, are endowed with certain fundamental rights that it is wrong not to respect. According to the evolutionary scenario set out by Darwin, the contemporary moral ideal of human rights is a response to a perception, however vague and indefinite, that mankind worldwide is united into one society, one community, however indeterminate or yet institutionally unorganized. As Darwin presciently wrote:

> As man advances in civilization, and small tribes are united into larger communities, the simplest reason would tell each individual that he ought to extend his social instincts and sympathies to all the members of the same nation, though personally unknown to him. This point being once reached, there is only an artificial barrier to prevent his sympathies extending to the men of all nations and races. If, indeed, such men are separated from him by great differences of appearance or habits, experience unfortunately shows us how long it is, before we look at them as our fellow-creatures (*4*).

The Biotic Community

Although 'The Ethical Sequence' of Leopold's "The Land Ethic" is clearly a compact summary of Darwin's more protracted account, Leopold parts company with Darwin in the way he envisions the next step in the sequence, the step beyond a universal human rights ethic. Darwin went on to say immediately what he foresaw:

> Sympathy beyond the confines of man, that is, humanity to the lower animals, seems to be one of the latest moral acquisitions.... This virtue, one of the noblest with which man is endowed, seems to arise incidentally from our sympathies becoming more tender and more widely diffused until they are extended to all sentient beings (*4*).

Darwin, as we see, projects the ethical sequence beyond a universal human rights ethic to what is today known as animal liberation/ rights or animal welfare ethics. As I have argued at some length and in a variety of venues, animal welfare ethics and Leopold's land ethic are not only theoretically divergent, they are pragmatically contradictory (*1*). Leopold never suggested that sentiency bears any moral significance. And the integrity, stability, and beauty of the biotic community and the welfare of sentient animals are often in direct conflict.

Furthermore, Darwin regards sympathy to the lower animals to be *incidental*, to be sort of a spill-over effect from our sentiments becoming "more tender and more widely diffused." The land ethic, on the other hand, requires "respect [if not sympathy] for...fellow members [of the biotic community] and also respect for the community as such" (*11*). Fellow-members may be plants or even soils and waters as well as sentient animals. As Leopold says, such respect is "implied" by a land ethic. It is not merely incidental.

The additional conceptual element in Leopold's argument, which alters the thrust of the ethical sequence and *logically* compels the extension of ethical criteria beyond the confines of humans, is the ecological notion of a "biotic community"—which is, of course, altogether absent in Darwin's discussion. (Darwin may be excused for not having explored the moral implications of the biotic community or seen beyond the incidental extension of human sympathies to sentient beings for the simple reason that modern ecology from which Leopold draws the concept of the "biotic community" was dependent upon Darwin's theory of evolution for its development!)

As the foreword to *A Sand County Almanac* makes plain, the overarching thematic principle of the book is the inculcation of the idea, through narrative description, discursive exposition, abstractive generalization, and occasional preachment, "that land is a community." The concept of a biotic community is "the basic concept of ecology" (*11*). Once land is popularly perceived as it is professionally perceived in ecology—as an integrated community—a correlative land ethic will emerge in the collective cultural consciousness.

According to Leopold, the next step in the Darwinian ethical sequence beyond the still incomplete ethic of universal humanity, but clearly discernible on the horizon, is not animal welfare ethics but the land ethic. The "community concept" has, so far, propelled the development of ethics from the savage clan to the family of humankind. "The land ethic simply enlarges the boundary of the community to include soils, waters, plants, and animals, or collectively: the land" (*11*).

The concept of a "biotic community" was developed as a working model or paradigm for ecology by Charles Elton in the 1920s (*27*). The natural world is organized as an intricate corporate society in which plants and animals occupy "niches" or, as Elton alternatively called them, "roles" or "professions" in the economy of nature (*6*). As in a feudal community, little or no socioeconomic mobility (upward or otherwise) exists in the biotic community. One is born to one's trade.

Leopold, in effect, argues that just as human society is founded in large part upon mutual security and economic interdependency and preserved only by limitations on freedom of action in the struggle for existence, that is, by ethical constraints, the biotic community that exhibits, as modern ecology reveals, an analogous structure can only be preserved, given the newly amplified impact of "mechanized man," by analogous limitations on freedom of action, that is, by a land ethic. A land ethic, furthermore, is not only "an ecological necessity" but an "evolutionary possibility" because a moral response to the natural environment — Darwin's social sympathies, sentiments, and instincts translated and codified into a body of principles and precepts — would be automatically triggered in human beings by ecology's social representation of nature (11).

Therefore, the key to the emergence of a land ethic is, simply, universal ecological literacy.

Three Scientific Cornerstones

The land ethic rests upon three scientific cornerstones: (1) evolutionary and (2) ecological biology set in a background of (3) Copernican astronomy. Evolutionary theory provides the conceptual link between ethics and social organization and development. It provides a sense of "kinship with fellow-creatures" as well — "fellow-voyagers" with us in the "odyssey of evolution" (11). It establishes a diachronic link between people and nonhuman nature.

Ecological theory provides a synchronic link, the community concept, a sense of social integration of human and nonhuman nature. Human beings, plants, animals, soils, and waters are "all interlocked in one humming community of cooperations and competitions, one biota" (12). The simplest reason, to paraphrase Darwin, should, therefore, tell each individual that he or she ought to extend his or her social instincts and sympathies to all members of the biotic community though different from him or her in appearance or habits.

Although Leopold never mentions it in *A Sand County Almanac,* the Copernican perspective, the perception of the Earth as "a small planet" in an immense and utterly inhospitable universe beyond, contributes, perhaps subconsciously but nevertheless powerfully, to our sense of kinship, community, and interdependence with fellow denizens of the Earth. It scales the Earth down to something like a cozy island paradise in a desert ocean.

Here in outline, then, are the conceptual and logical foundations of the land ethic: Its conceptual elements are a Copernican cosmology, a Darwinian proto-sociobiological natural history of ethics, Darwinian ties of kinship among all forms of life on Earth, and an Eltonian model of the structure of biocoenoses all overlaid on a Humean-Smithian moral psychology. Its logic is that natural selection has endowed human beings with an affective moral response to perceived bonds of kinship and community membership and identity; that today the natural environment, the land, is represented as a community, the biotic community; and that, therefore, an environmental or land ethic is both possible — the bio-psychological and cognitive conditions are in place — and necessary because human beings collectively have acquired the power to destroy the integrity, diversity, and stability of the environing and supporting economy of nature.

Ethical Holism

The land ethic is unique among its precursors in another important feature — its provision of what Kenneth Goodpaster carefully called "moral considerability" for the biotic community *per se*, not just for fellow members of the biotic community (7). The land ethic has, in other words, a holistic as well as an individualistic cast.

Indeed, as "The Land Ethic" develops, the focus of moral concern shifts gradually away from plants, animals, soils, and waters individually to the biotic community collectively. Toward the middle, in 'Substitutes for a Land Ethic,' Leopold invokes the "biotic rights" of *species*, as the context indicates, of wildflowers, song birds, and predators. In 'The Outlook,' the last climactic section of "The Land Ethic," nonhuman natural entities, first appearing as fellow members, then considered in profile as species, are not so much as mentioned in what might be called the "summary moral maxim" of the land ethic: "[A] thing is right when it tends to preserve the integrity, stability, and beauty of the biotic community. It is wrong when it tends otherwise" (*11*).

Ecological thought, historically, has tended to be holistic in outlook (*27*). Ecology is the study of the *relationships* of organisms to one another and to the elemental environment. These relationships bind the *relata* — plants, animals, soils, and waters — into a seamless fabric. The ontological primacy of objects and the ontological subordination of relationships, characteristic of classical Western science, is, in fact, reversed in ecology (*3*). Ecological relationships determine the nature of organisms rather than the other way around. A species is what it is, in other words, because it has adapted to a *niche* in the ecosystem. The whole, the system itself, thus literally and straightforwardly shapes and forms its component species.

In 1935 British ecologist Arthur Tansley pointed out that from the perspective of physics the "currency" of the "economy of nature" is energy (*25*). Tansley suggested that Elton's qualitative and descriptive food chains, food webs, trophic niches, and biosocial professions could be quantitatively expressed by a thermodynamic flow model. It is Tansley's state-of-the-art thermodynamic paradigm of the environment that Leopold explicitly sets out as a "mental image of land" in relation to which "we can be ethical" (*11*).

'The Land Pyramid' is the pivotal section of "The Land Ethic," the section that effects a complete transition from concern for "fellow-members" to concern for the "community as such." It is also its longest, most technical section. A description of the "ecosystem" (Tansley's deliberately nonmetaphorical term) begins with the sun. Solar energy "flows through a circuit called the biota" (*11*). It enters the biota through leaves of green plants and courses through plant-eating animals, then on to omnivores and carnivores. At last the tiny fraction of solar energy converted to biomass by green plants remaining in the corpse of a predator, animal feces, plant detritus, or other dead organic material is garnered by decomposers—worms, fungi, bacteria. They recycle the elements and degrade into entropic equilibrium any remaining energy. According to this paradigm:

> Land, then, is not merely soil; it is a fountain of energy flowing through a circuit of soils, plants, and animals. Food chains are the living channels which conduct energy upward; death and decay return it to the soil. The circuit is not closed; ...but it is a sustained circuit like a slowly augmented revolving fund of life.... (*11*).

In this exceedingly abstract (albeit poetically expressed) model of nature, process precedes substance and energy is more fundamental than matter. Individual plants and animals become less autonomous beings than ephemeral structures in a patterned flux of energy. Maintenance of "the complex structure of the land and its smooth functioning as an energy unit" emerges in 'The Biotic Pyramid' as the *summum bonum* of the land ethic (*11*).

Eating and Being Eaten

The living channels, "food chains," through which energy flows, are composed of individual plants and animals. A central, stark fact lies at the heart of ecological processes: energy, the currency of the economy of nature, passes from one organism to another, not from hand to hand, like coined money, but, so to speak, from stomach to stomach. Eating *and being eaten*, living *and dying* are what makes the biotic community hum.

The land ethic, like all previous stages in the ethical sequence, reflects and reinforces the structure of the community to which it is correlative. Trophic asymmetries constitute the kernel of the biotic community. It seems unjust, unfair. But that is how the economy of nature is organized (and has been for thousands of millions of years). A "right to life" for individual members is not consistent with the structure of the biotic community and, hence, is not mandated by the land ethic.

However, our recognition of the biotic community and our immersion in it does not imply that we do not also remain members of the human community — the "family of man" or "global village" — or that we are relieved of the attendant and correlative moral responsibilities of that membership — among them to respect universal *human rights* and uphold the principles of individual *human* worth and dignity. The biosocial development of morality does not grow in extent like an expanding balloon, leaving no trace of its previous boundaries, so much as like the circumference of a tree (*18, 23*). Each emergent, larger social unit is layered over the more primitive, intimate ones. Hence, family obligations come before nationalistic duties and humanitarian obligations come before environmental duties. The land ethic, thus, does not cancel human morality.

Respect for Fellow Members

Neither is it inhumane. Nonhuman members of the biotic community have no "human rights" because they are not, by definition, members of the human community. According to Leopold, they do have a "biotic right" *as species* to continued existence. And *individual* plants and animals, as fellow members of the biotic community, deserve, according to Leopold, respect.

How exactly to express or manifest respect, while at the same time abandoning our fellow members of the biotic community to their several fates or even actively consuming them for our own needs (and wants), or deliberately making them casualties of wildlife management for ecological integrity, is a difficult and delicate question.

Fortunately, American Indian and other traditional patterns of human-nature interaction provide rich and detailed models. Algonkian woodland peoples, for instance, represented animals, plants, birds, waters, and minerals as other-than-human persons engaged in reciprocal, mutually beneficial socioeconomic intercourse with human beings (*17*). Tokens of payment, together with expressions of apology, were routinely offered to the beings whom it was necessary for these Indians to exploit. Care not to waste the usable parts, and in the disposal of unusable animal and plant remains, were also an aspect of the respectful, albeit necessarily consumptive, Algonkian relationship with fellow members of the land community. As I have more fully argued elsewhere, the Algonkian portrayal of human-nature relationships is, indeed, although certainly different in specifics, identical in abstract form to that recommended by Leopold in "The Land Ethic" (*2*).

Leopold to my knowledge never consciously patterned his land ethic on an American Indian paradigm. The correspondence between the Leopold land ethic and the Algonkian land ethic springs from the essentially social representation of nature common to both. The myths and stories of the woodland peoples of the western Great Lakes portrayed animals and plants as "persons" engaged in socioeconomic intercourse with human beings (*8*). The theory of ecology, upon which Leopold relies for his understanding of nature, carefully avoids anthropomorphic personification, but the outcome is formally the same. Human beings are entwined in a symbiotic community that includes plants and animals, soils and waters, as well as persons proper. The Leopold land ethic, like the Algonkian land ethic, acknowledges and affirms our biotic citizenship. The difference is that the Indians expressed and affirmed their participation in the biotic community in the language of myth and legend—as befits an oral culture—while Aldo Leopold expressed and affirmed the same idea in the abstract language of evolutionary and ecological biology. And that was his special contribution to the articulation of a credible and convincing environmental ethic in the Age of Ecology.

REFERENCES

1. Callicott, J. Baird. 1980. *Animal liberation: A triangular affair*. Environmental Ethics 2: 311-338.
2. Callicott, J. Baird. 1982. *Traditional American Indian and western European attitudes toward nature: An overview*. Environmental Ethics 4: 163-174.
3. Callicott, J. Baird. 1986. *The metaphysical implications of ecology*. Environmental Ethics 8: 300-315.
4. Darwin, Charles R. 1904. *The descent of man and selection in relation to sex*. J. A. Hill and Company, New York, New York.
5. Deutsch, Eliot. 1986. *A metaphysical grounding for natural reverence*. Environmental Ethics 4: 293-299.
6. Elton, Charles. 1927. *Animal ecology*. Macmillan, New York, New York.
7. Goodpaster, Kenneth. 1978. *On being morally considerable*. The Journal of Philosophy 22: 308-325.
8. Hallowell, A. Irving. 1960. *Ojibwa ontology, behavior, and world view*. In S. Diamond [editor] *Culture in History: Essays in Honor of Paul Radin*. Columbia University Press, New York, New York.
9. Hamilton, W. D. 1964. *The genetical theory of social behavior*. Journal of Theoretical Biology 7: 1-52.
10. Hume, David. 1777. *An inquiry concerning the principles of morals*. The Clarendon Press, Oxford, England.
11. Leopold, Aldo. 1949. *A Sand County almanac and sketches here and there*. Oxford University Press, New York, New York.
12. Leopold, Aldo. 1953. *Round River*. Oxford University Press, New York, New York.
13. Muir, John. 1901. *Our national parks*. Houghton Mifflin Co., Boston, Massachusetts.
14. Muir, John. 1916. *A thousand mile walk to the Gulf*. Houghton Mifflin Co., Boston, Massachusetts.
15. Nash, Roderick. 1982. *Wilderness and the American mind*. Yale University Press, New Haven, Connecticut.
16. Nash, Roderick. 1987. *Aldo Leopold's intellectual heritage*. In J. Baird Callicott [editor] *Companion to a Sand County Almanac*. University of Wisconsin Press, Madison.
17. Overholt, Thomas W., and J. Baird Callicott. 1982. *Clothed-in-fur and other tales: An introduction to an Ojibwa world view*. University Press of America, Washington, D.C.
18. Routley, Richard, and Val Routley. 1980. *Human chauvinism and environmental ethics*. In D. Mannison, M. McRobbie, and R. Routley [editors] *Environmental Philosophy*. Department of Philosophy, Research School of the Social Sciences, Australian National University, Canberra.
19. Sahlins, Marshall. 1972. *Stone age economics*. Aldine Atherton, Chicago, Illinois.
20. Schopenhauer, Arthur. 1961. *The world as will and idea*. Doubleday & Co.,

Garden City, New York.

21. Schweitzer, Albert. 1976. *The ethic of reverence for life*. In Tom Regan and Peter Singer [editors] *Animal Rights and Human Obligations*. Prentice Hall, Inc., Englewood Cliffs, New Jersey.

22. Service, Elman R. 1962. *Primitive social organization: An evolutionary perspective*. Random House, New York, New York.

23. Singer, Peter. 1983. *The expanding circle: Ethics and sociobiology*. Farrar, Straus, and Giroux, New York, New York.

24. Smith, Adam. 1759. *Theory of the moral sentiments*. A. Millar, A. Kinkaid, and J. Bell, London, England.

25. Tansley, Arthur. 1935. *The use and abuse of vegetational concepts and terms*. Ecology 16: 292-303.

26. Wilson, Edward O. 1975. *Sociobiology: The new synthesis*. Harvard University Press, Cambridge, Massachusetts.

27. Worster, Donald. 1977. *Nature's economy: The roots of ecology*. Sierra Club Books, San Francisco, California.

II

Standing on His Shoulders

At his desk, 424 University Farm Place, late 1930s

6

The Land Ethic
and the World Scene

Raymond F. Dasmann

I did not have the privilege of knowing Aldo Leopold. But during my undergraduate education, under the principal tutelage of his son Starker Leopold, I was introduced to what had become the standard textbook in wildlife courses, Aldo Leopold's *Game Management*. It was more than a textbook; it was a beautifully written volume that skillfully presented the knowledge of wildlife management available at that time. More than any other single contribution, it established the field of wildlife management as both a science and an art. Then, during my first year as a graduate student, *A Sand County Almanac* was published. In this historic book, Aldo Leopold reexamined our code of ethics and extended its rules to the nonhuman world. Both books made a powerful impression on my own thinking.

I cannot pretend to expertise in philosophy, theology, or the world scene. Hence, I must ask readers to bear with me as I present my view of the world scene, based on limited sampling, and my ideas of Leopold's contribution, based on no knowledge of the man himself but on a great respect for his work and his ideas.

When I again read *Game Management* recently, I was impressed with its current relevance. It is not outdated. Admittedly, we have more information, greater sophistication in the manipulation of information, and more ecological theories with which to play. But most of Leopold's basic concepts remain sound and useful.

But what struck me most was a sense of nostalgia. In many ways

107

America was a much nicer place to live when Leopold wrote *Game Management*, 50 or 60 years ago. There were only half as many Americans and wildlands were far more extensive than today. Americans, as a people, tended to be more secure and confident, despite the economic disruption caused by the Depression and the Dust Bowl. We lacked much in the way of military trappings, but we did not feel personally threatened by the blusterings of a Hitler or the militarism of Japan. We felt no obligation to take care of the world's environment. It did not seem to need our care. In fact, the natural world, as yet little modified by human activity, was so extensive that the need for conservation and management was apparent only to the farsighted and best informed. Vast areas of tropical forest had only felt the bite of a few primitive axes. Great expanses of savanna and steppe experienced mostly the pounding of wild hooves. High mountain ranges were still remote and mostly empty of people. Wildlife spectacles, the great herds and flocks, were still to be seen in much of Africa, Latin America, India, China, the East Indies, even Iran and Arabia. Except in a few areas of concentration, there were relatively few people on Earth.

Despite this picture of overall biotic luxuriance, in those countries that had been colonized fully by Europeans — the United States, Canada, Australia, New Zealand, South Africa — wildlife populations were at a low ebb. When Aldo Leopold was born, Americans had just about finished off the bison as a species, eliminating tens of millions of animals from the plains and prairies. During his growing years, we drastically reduced the populations of most large animals. Wolves, mountain lions, the grizzly bear, elk, moose, mountain sheep, mountain goats, and pronghorn seemed to be on their way out.

Thanks to the new concepts of game management, particularly Leopold's emphasis on the overriding importance of habitat, Americans have restored, to some extent, most of those depleted populations. Overall, wildlife management in the United States has been a success story, at least until recently. The same is true in Canada, Australia, South Africa, and New Zealand.

I cannot, of course, credit one person for the spread of these new wildlife conservation ideas. Nevertheless, Leopold's blending of the ancient art of wildlife husbandry, gained from the experience of the gamekeepers on the private estates of Europe, with the scientific concepts being acquired by biologists, ecologists, and others, and the application of this approach to the wildlands that had been taken over by white settlers, certainly was an important contribution to this success.

I hardly need to say that the world situation has changed in the past 50 years. The Earth's population has more than doubled. Economic demand continues to grow. Too many people believe the good life can be bought. The means of exploiting resources and destroying resource capital have become enormously sophisticated. Everybody seems to have a chain saw, and bulldozers are more prominent than one cares to think. Globally, a pall of pollution threatens even those areas still untouched by mechanical forms of exploitation. We are faced with a possible loss of wild species of cataclysmic proportions, equivalent to that brought about during the extinction of the dinosaurs and related life forms at the beginning of the Cenozoic Era.

Furthermore, while earlier extinctions were caused by external or unavoidable forces, the present potential for catastrophe is entirely the result of human action and is for the most part avoidable. Worst of all, we seem paralyzed and devoid of will in the face of this creeping catastrophe. Our only response is to call for more of the same things that have contributed to it. I am reminded of a cartoon I saw some years ago. If all the king's horses and all the king's men can't put humpty-dumpty together again, then obviously the king needs more horses and more men. Right? This has been our approach. Isn't it time to try something new? If so, perhaps we should first look at what Leopold wrote in *A Sand County Almanac:*

> A system of conservation based solely on economic self-interest is hopelessly lopsided. It tends to ignore, and thus eventually to eliminate, many elements in the land community that lack commercial value, but that are (as far as we know) essential to its healthy functioning. It assumes, falsely, I think, that the economic parts of the biotic clock will function without the uneconomic parts. It tends to relegate to government many functions eventually too large, too complex, or too widely dispersed to be performed by government. An ethical obligation on the part of the private owner is the only visible remedy for these situations.

And elsewhere Leopold wrote:

> No important change in ethics was ever accomplished without an internal change in our intellectual emphasis, loyalties, affections,

and convictions. The proof that conservation has not yet touched these foundations of conduct lies in the fact that philosophy and religion have not yet heard of it. In our attempt to make conservation easy, we have made it trivial.

Focusing on one of the growing problems that threaten the biosphere, the loss of biotic diversity, Leopold's statement could have been written today. There is no justification, in economic terms, for the conservation of a million or more species of tropical insects, most of which have yet to be baptized by science and given a scientific name. Yet most entomologists and ecologists believe, although they cannot prove, that these insects are essential to the functioning of the biological clocks of tropical forest ecosystems. It is difficult enough for ecologists to justify that tropical forest systems in their entirety are important to human survival, despite all of the obvious benefits that we derive from them. It is much more difficult to make a case for individual species within those forests. Faced with a development scheme that supposedly would bring hundreds of millions of dollars into public or private coffers, the arguments for saving even hundreds or thousands of species do not seem to weigh much in an economic cost-benefit analysis.

Unfortunately, it is even difficult to prove the economic importance of the large and colorful species that many of us admire, those that Michael Soule has termed the metacharismatic megavertebrates. The condor has virtually disappeared from the California sky and perhaps may never be restored. Have the California ecosystems that the condor once touched suffered even minor perturbations? Probably not. The grizzly bear has disappeared from most of the United States. Did its demise cause any economic loss worth mentioning? Again, probably not. Nevertheless, people will vote and put up money to save condors and grizzy bears. The U.S. Endangered Species Act is supported by members of both political parties. Enough Americans have a strong enough ethical obligation to, aesthetic value for, or feelings of identity with wild creatures to want to maintain wild America. The conservation movement has grown strong in this country, not because of economic arguments about wildlife, but because millions of Americans—most of us, according to the polls—are willing to forego some level of economic benefit to make sure that wildlife survives. Here, at least, philosophy and religion are becoming aware of conservation and Leopold's ethic. But this point of view is not necessarily present or prevalent in the developing countries of the world, where the choice often seems to be one of wildlife or human survival.

To save biological diversity, we must look to the countries of the tropical world. That is where most of the diversity is to be found and is most endangered. The term biological diversity includes genetic diversity, gene pools, germ plasm resources, and much more. It is a new term for nature, which, as a word, has lost some of its glamor. Everyone was for nature, just as they were for motherhood. But getting appropriations to save nature was becoming just as difficult as getting appropriations to make motherhood easier. Give nature a new, hard-edged, scientific name — biological diversity — and the situation changes. Congress recently passed a bill that would set aside $10 million in U.S. Agency for International Development funds each year to save biological diversity in other countries. Of course, $10 million will only buy more king's horses and more king's men for humpty-dumpty, unless we change our approach.

In *Game Management*, Leopold presented a chronology of the development of wildlife conservation in the United States. Paraphrased and condensed, the sequence of events was (1) restriction of hunting, (2) control of predators, (3) reservation of land as national parks and game reserves, (4) artificial restocking through captive breeding, (5) starting education work, (6) starting wildlife research, and (7) improving habitat through public and private management.

In much of the tropical world there has been a similar sequence. First, laws are passed to control hunting, which are usually ineffective because most people do not hear about the laws and no one enforces them. Second, recognizing the impossibility of enforcing laws throughout the country, national parks or game reserves have been established. In these areas laws have a better chance of being enforced through the presence of wardens, rangers, and game scouts. Third, predators have been controlled to protect ungulates inside the parks and reserves and to protect humans and their livestock outside of parks. Fourth, to a limited extent, some countries have begun restocking species through transplants or captive breeding. Limited educational work has been done, usually in cities. Some research has been started. Finally, little or nothing has been done in the way of improving and managing wildlife habitat.

In many tropical countries this sequence of events had a further disadvantage. It was imposed originally from outside by colonial governments, which ignored the wishes and ideas of the local people. Reserves were established on the former homelands of tribal people, and the people were moved out, usually with little or no compensation or consideration. Old ways of living, based often on a sustainable use of wildlife, were brought to a halt. Hunting was restricted to the colonial masters, with some consideration for the local elite. Most local people could not even visit national parks. It is little wonder that these people had no great enthusiasm for wildlife conservation.

Yet these same local people had created or maintained the conditions, through sustainable use of wild resources, that caused the outsiders to decide that these areas were suitable for national park status because of the relative abundance of wildlife. Local people often had behavioral restraints — call them ethics, although they were often not at a level of individual awareness — that prevented the destruction of habitat and wildlife. However, these ethics were set aside by government, and local people were left with only one role in relation to wildlife, that of poacher and lawbreaker.

When foreign rule ended and new governments took over independent nations, often there were few changes in relation to wildlife. The new leaders mostly were educated overseas and preferred the advice of experts to the opinions of local people. They saw national parks and wildlife as national assets, bringing tourists and their money. The idea of sharing responsibility for wildlife with the rural population did not appeal to those leaders who preferred to rule from the top down.

Unfortunately, only part of Leopold's message has come through in most countries — protection or preservation. This may be because the European and American conservationists who promoted conservation ideas in the tropical world most often were of a preservationist persuasion. Wildlife managers were busy counting ducks or deer at home. The idea that wildlife populations are dynamic and can sustain cropping by humans was accepted only on a limited basis — enough to permit some safari hunting by the elite. The idea that protection will not save wildlife unless suitable habitat is maintained, and that this may require active management, also was only partially accepted. Leopold's overriding idea that the only person who can effectively manage wildlife is the person who controls the use of the land — the landowner or land manager — did not seem to come through at all. Hence, there is a one-sided approach to conservation that, in the long run, probably will fail.

Fortunately, this situation is changing in some countries. Zimbabwe, for example, has placed the responsibility for wildlife in the hands of local people through local conservation councils. The local economy benefits directly from the sale of hunting privileges and from meat and other wildlife products produced. One trophy bull elephant, with hunting rights sold by competitive bid, can bring $24,000, much of which goes to the local economy. This encourages protection and management of those animals that once were seen as destructive nuisances. Zimbabwe's wildlife department serves only as an overseer, checking to see that endangered species are protected and that cropping levels are balanced by productivity. However, only a few countries are taking such steps.

If efforts to conserve biotic diversity are to succeed, nature conservation must become a part of the total land use pattern. There must be places for wild species in cities and on farms, on rangelands and in forests, and in the nature reserves and wildernesses. This cannot be attained through a preservationist stance. Sustainable use and management must have a role. This is a Leopold doctrine. It is needed today more than ever.

This finally brings me to the subject of ethics. Leopold argued very persuasively for the need for an extension of ethics from man to the land. He argued that we had developed an ethical code guiding human relationships. We needed a similar code guiding our relationships with nature. We must develop, he said, an ecological conscience at least as strong as that guiding our activities toward fellow humans. For this to happen, we first needed an "internal change in our intellectual emphasis, loyalties, affections, and convictions."

I believe we are moving toward such an extension of ethics in this country. Progress toward it, at least as reflected in the biased sample of people that I deal with, has been impressive. Obviously there is a long way to go.

Other countries have had ethical codes concerning animal life. In southern Asia, Hindus and Buddhists have advocated the doctrine of ahimsa, of doing no harm to other beings. Some fundamentalists have extended this protection even to flies, fleas, and mosquitoes. Unfortunately, the doctrine usually has not been applied to the plant world—even the devout must eat—so habitat destruction has gone on without apparent realization that this was only an indirect means of killing. Furthermore, the killing of animals also has occurred when the financial incentive was strong enough. The elephant may be the representative on Earth of a sacred being. It still gets blasted when it invades rice paddies. Regardless of their religion, most people seem to share the weakness of most Christians — religious ethics are regarded piously on holy days, but set on the shelf the rest of the week.

I agree with Aldo Leopold that we need an extension of ethics, not just from man to the land but to guide our relationships with all other human beings. Our inhumanity to each other establishes the basis for our behavior to other species. Our willingness to oppress fellow humans to enhance our own economic well-being makes it more certain that they in turn will eliminate wildlife to provide for their own economic survival. We cannot extend our concern for wild nature unless we are also concerned for the welfare of people because the two cannot be separated. Where the basic needs of people are not met, not even the strongest military can enforce game laws or protect nature reserves for long.

A decade or two ago these words would not have been received with much enthusiasm. But a lot has changed recently. The human dimension is being considered in projects that not long ago would have been oriented entirely toward nature conservation. World Wildlife Fund projects take into account the need to address human needs as well as wildlife needs. Even the World Bank now considers the needs of tribal people and the natural environments in which they live.

I do not believe Aldo Leopold addressed these issues fully because he had not seen the onslaught on the natural world and our fellow humans that has occurred since 1948. He also was influenced to believe that people are nicer than they really are. Nevertheless, I see what can be called creeping Leopoldism spreading in the world. I have hope that the future will witness not only an extension of our ethics but a willingness to live by them.

7

North American Deer Ecology: Fifty Years Later[1]

Dale McCullough

Throughout his career, Aldo Leopold was probably more involved with deer than any other wildlife species. His early career with the Forest Service in the Southwest was largely directed to understanding deer ecology in the Gila and Kaibab National Forests. His conservation efforts focused on organizing sportsmen to protect deer and foster their increase. He then moved to Wisconsin where much of his time was devoted to comparing the deer situation in northern forests of the Lake States with that in the Southwest. His final and most turbulent years were spent as a Wisconsin state conservation commissioner fighting hunter opposition to reductions in overpopulations of deer in Wisconsin. His assault on the "citadel of the sacred doe," coming so soon after it was built and fortified, was an undertaking worthy of Indiana Jones.

We are fortunate that Aldo was a prolific writer. From his publications, letters, and personal notes one can track his thinking throughout his career (3). It is reassuring to find that he changed his mind about things—sometimes changing it back again—and that he struggled with uncertainties in interpretation and was frustrated by the lack of research results. These are the same problems we confront today, and we can only hope that we face the challenge as resolutely as Leopold did.

There are two major areas in which I will compare deer ecology then and now:

---

[1]This chapter is based on the 23rd annual Paul L. Errington Memorial Lecture.

115

1. Population concepts or models, that is, how we think deer populations work.

2. Translation of population concepts to real-world deer management programs.

Population Concepts or Models

Early in his work in the Southwest, Leopold derived a concept of wildlife populations that envisioned two opposing factors: (1) breeding potential, an innate drive of a population to increase at an exponential rate, and (2) environmental resistance, the sum total of the environmental factors opposed to this increase that results in a dynamic equilibrium of animal numbers. The population at the equilibrium Leopold designated as carrying capacity (5).

This word "model" was sufficiently robust to incorporate the important concepts of Paul Errington about thresholds of security (numbers below carrying capacity are quite immune to mortality factors, while those above are extremely vulnerable) and compensation (equivalent to density-dependence) (2). This model is the predominant concept of population dynamics still current in wildlife management.

I tested these concepts of population dynamics by experimental manipulation of the white-tailed deer herd on the Edwin S. George Reserve in southwestern Michigan. The George Reserve is a 1,146-acre, deer-fenced area near Pinckney, owned by the University of Michigan. The deer herd originated from a transplant of six animals in 1928. An annual drive census began in 1933 and, with the exception of a few years, has been conducted annually to date. My studies began in 1966 and continue.

The university holds a game breeder's license that allows the herd to be managed without regard for state hunting regulations. Any number of deer of any sex or age can be removed at any time by any means. This freedom allowed experimental manipulation that is seldom possible. Manipulations of population were designed to test density-dependent responses and to elucidate the relationship between vegetation and deer. Details of this research have been published elsewhere (6, 7, 8).

Results of these experiments broadly supported the ideas of Leopold and Errington. The population showed density-dependent changes in birth and death schedules, hence, the "compensation" addressed by Errington. The population showed a high growth rate when beginning from a small, initial population, illustrating the validity of Leopold's idea of "breeding potential." The growth rate slowed as the capacity of the reserve to support deer was reached, an expression of Leopold's idea of "environmental resistance." The population also showed a tendency to exceed carrying capacity, Leopold's "irruptive population response." In sum, the George Reserve deer population experiment confirmed the concepts of Leopold and Errington about population dynamics.

Inherent in the Leopold population model is a corresponding "theory of harvest" model. Curiously, neither Leopold nor Errington seemed to have recognized this fact. Instead, Leopold independently derived a harvest theory based upon the idea of a "harvestable surplus." The population was viewed as producing an annual oversupply that could be removed by sport hunting without reducing the breeding population in the following season. This led to a passive application of harvesting in that the "surplus" was measured before determining the allowable kill.

Modern harvest theory and the George Reserve deer population studies show that harvest can be used to drive the population to a more productive state. Because of the density-dependent response, populations reduced well below carrying capacity show increased recruitment of young individuals to replace harvest removals. Thus, populations initially near carrying capacity can be harvested at greater than the current "harvestable surplus" and, so long as the maximum sustainable harvest is not exceeded, the subsequent increased recruitment replaces the harvest removal.

Although Errington was one of the original architects of the concept of compensation, there is little evidence that he recognized the harvest implications of the concept. The record is clear that Leopold was unaware of these possibilities. While he recognized that large harvests were required to reduce excessive deer numbers, he saw them as redress of a crisis situation rather than as a source of continued, high sustained harvest. This is unfortunate because high sustainable harvests from lower populations could be obtained simultaneously with protection of environmental quality. It was this latter goal that he strived so valiantly to achieve through his essays on land ethics and with protection of vegetation communities from too many hungry deer.

Leopold and Errington can hardly be faulted for failing to recognize these relationships. The wildlife profession has proven remarkably impervious to harvest theory. William Ricker, who had been exploring these ideas in fisheries, presented them in the *Journal of Wildlife Management* in 1954 (*10*), then R. F. Scott independently presented them at the North American Wildlife Conference the same year (*12*). Neither paper seems to have had any impact. The same ideas were presented later by several researchers, all more or less independently (*1, 4, 6, 11, 13*).

I treated harvest theory extensively in my 1979 book on the George Reserve deer herd. Many who read the book believed my treatment of harvesting to be a radical departure. This surprised me because the harvest theory I put forward was exactly that predicted by the Leopold population model. I thought I was one of the true believers preaching that old-time religion. I hadn't anticipated the long delay in acceptance and implementation of harvest theory in wildlife management.

Textbooks are written by academics at the forward edge of knowledge. Of five wildlife management textbooks published since 1984, three completely omit harvest theory, one treats it passingly, and one presents some of the ramifications for setting harvest policy. Many workers in the wildlife field are unaware of harvest theory, which is remarkable given that setting harvest regulations is one of the major responsibilities of state and federal wildlife agencies.

Deer Management in the Real World

Aldo Leopold did not live long enough to see antlerless-deer hunting as a routine management practice in Wisconsin. Nevertheless, his resolute pursuit of this goal, even when he was the sole spokesman for this unpopular position on the Wisconsin Conservation Commission, helped greatly with its eventual acceptance, which, ironically, came the year after his death. Many states now routinely have antlerless-deer seasons. Others are still back in Leopold's time, with exclusively bucks-only hunting. California, which prides itself in setting opinion for the rest of the nation, is only now showing signs of emerging from the dark ages in deer management. Even in the most progressive states there is a large contingent of hunters ready to join ranks with any self-proclaimed leader who announces a fight against the heresy of the state wildlife department that persists in treating does and fawns as fit game for real sportsmen. This latent opposition suggests that there must be some substantial root cause of the perennial resistance to anterless-deer hunting.

One consequence of high sustainable harvests of large mammals, such as deer, is the inevitable lowering of animal numbers below the numbers present under light harvests typical of bucks-only seasons. This relationship between harvest size and overall population is not understood by many wildlife professionals, and it becomes a major stumbling block in communication between hunters and deer biologists.

A hunter's assessment of the deer population is based on the number of deer and deer sign seen while hunting. If the hunter sees many deer and much sign, he or she will be satisfied with the status of the herd, even though the hunter may not have gotten a deer. Conversely the hunter will be concerned if he or she sees few deer and sign, even if the hunter manages to take a large trophy animal.

The biologist, on the other hand, is concerned about high sustainable harvests, which he or she assumes will result in satisfied hunters because many more will go home with a deer. Ironically, the management program that puts the most deer in the freezer is the program that results in fewer deer in the field. The consequence is that program success by the biologist's criterion results in failure by the hunter's criterion, and vice versa.

Until this communication gap is closed and the inverse relationship between harvest and deer in the field is recognized by both parties, the adversarial relationship between hunters and wildlife managers is likely to continue.

In recent years hunters and biologists have been putting aside their differences to confront a mutual outside threat. The shotgun that forced this marriage of convenience was the spectacular growth and influence of protectionism. This movement seeks to ban all blood sports, among which hunting is a particularly inviting target. To hunters the prospect that anti-hunting sentiment might prevail may seem like the end of the world. But from an ecological perspective the elimination of hunting would not be a disaster to the environment in most cases. The notable exception is overpopulation of large herbivores that destroy their own habitat. Deer are the most prevalent examples of overpopulation, with examples extending from Florida to Minnesota and Massachusetts to California. Newspapers continually report cases where overpopulations of deer are resulting in controversy. Protectionists have proposed a number of unique solutions to these problems.

The Angel Island Fiasco

One case involves black-tailed deer in Angel Island State Park, a one-square-mile island in San Francisco Bay. Deer are protected from hunting by park status, and there are no effective predators on the island. The deer population has built to starvation levels, with consequent habitat damage, three times in the recent past. The first peak occurred in 1966 when an estimated 300 deer were present on the island. Park rangers began shooting the animals to reduce the population. After 50 deer were taken, a public outcry forced discontinuance of the program. While alternatives were being argued in the media, the deer population crashed because of starvation.

In 1976 a second peak in deer numbers was reached, with a repeat of the previous conflict over shooting. The San Francisco Society for the Prevention of Cruelty to Animals sought and received permission from the State Parks and Recreation Department and the Department of Fish and Game to feed deer. There was debate about the outcome of the feeding program but, in any event, an estimated deer herd of 225 animals suffered a minimum of 100 known deaths. For a second time the population was reduced by starvation.

In 1980 the deer herd reached a third peak and was again at the starvation stage. Because shooting was not feasible, I proposed an introduction of coyotes to allow natural predation to reduce the herd. This proposal got wide media coverage and generated a storm of public outrage. The merits of the proposal were never considered in the emotional outburst against the "killer coyote" plan. My sanity and ancestry were much debated in newspaper letters to the editor and columns and on radio talk shows.

Eventually, the San Francisco Society for the Prevention of Cruelty to Animals forced the Parks and Recreation Department and the Fish and Game Department to relocate deer from Angel Island. A total of 203 deer were trapped and relocated to Cow Mountain, about 100 miles north of San Francisco. An additional 12 deer died or were sacrified during the capture effort, 16 were found dead on the island, and 44 were counted in a drive count following the relocation effort. Thus, a minimum of 275 deer had been on the island prior to the capture.

Mary O'Bryan, one of my graduate students, and I did a follow-up study of the relocated animals (9). We documented that about 85 percent of the relocated deer had died by the end of one year, most within three months after relocation. When these results were released to the public, it was the Society for the Prevention of Cruelty to Animals' turn for a media roasting. Although the Society steadfastly maintained that the program was a success, relocation was never again advanced as a solution to the deer problem.

It had become apparent to all that the deer population on Angel Island was destined to continue to build and crash. By 1983 the herd was again expanding. The Society for the Prevention of Cruelty to Animals proposed that the population be controlled by a trapping and sterilization program. The Parks and Recreation Department and the Fish and Game Department both gave permission for the Society to attempt the program. The Society undertook to treat females with antiovulatory drugs via subcutaneous implants. Preliminary calculations showed that at least 90 percent of the females (60 females) would have to be trapped and implanted to control further population growth. The capture effort began in August 1983 with much media coverage, anticipating trapping success like that in the Department of Fish and Game relocation program. Again, one of my graduate students, Gene Fowler, and I set out to document the outcome of the program.

The Society soon discovered that trapping deer in a rapidly expanding population was a much more difficult task than trapping deer when they were near starvation. After a prolonged effort, only 30 females had been trapped and treated. Following some agonizing deliberation, the Society quietly abandoned the program and further involvement in management of the Angel Island deer herd. Last year, without media attention, park personnel began shooting deer on Angel Island to limit the population.

In many ways the Angel Island deer controversy was a tempest in a teapot and a colossal waste of money. The outcomes of various treatments were exactly those predicted by deer biologists. But Angel Island is at the center of a large metropolitan area, and the deer problem generated enormous media attention. What better laboratory to demonstrate that alternatives to shooting for control of deer populations are expensive, ineffective, and not particularly humane. It was especially telling that an organization as prestigious, well-financed, and dedicated as San Francisco's Society for the Prevention of Cruelty to Animals could not find a workable alternative to shooting. The uncommitted public and media concluded that wildlife professionals knew what they were talking about while the protectionists did not. With luck, knowledge of the failure of alternative methods of deer population control on Angel Island will avoid their duplication elsewhere.

Harry Truman once said, "If you can't take the heat, stay out of the kitchen." For me, taking the heat on Angel Island was well worth the outcome. It felt good to be vindicated, of course; forgive me for indulging in a bit of glee at the discomfort of my colleagues with the Society for the Prevention of Cruelty to Animals. But the real gain was in the information obtained. How else could one conduct these unique experiments? I will always wonder, though, if those "killer coyotes" could have done the job.

REFERENCES

1. Caughley, G. 1977. *Analysis of vertebrate populations.* John Wiley and Sons, New York, New York. 234 pp.
2. Errington, P. L. 1967. *Of predation and life.* Iowa State University Press, Ames. 277 pp.
3. Flader, S. L. 1974. *Thinking like a mountain: Aldo Leopold and the evolution of an ecological attitude towards deer, wolves and forests.* University of Missouri Press, Columbia. 284 pp.
4. Gross, J. E. 1969. *Optimum yield in deer and elk populations.* Transactions of the North American Wildlife Conference 34: 372-385.
5. Leopold, Aldo. 1933. *Game management.* Charles Scribner's Sons, New York, New York. 481 pp.
6. McCullough, D. R. 1979. *The George Reserve deer herd: Population ecology of a K-selected species.* University of Michigan Press, Ann Arbor. 271 pp.
7. McCullough, D. R. 1982. *Population growth rate of the George Reserve deer herd.* Journal of Wildlife Management 46: 1,079-1,083.
8. McCullough, D. R. 1984. *Lessons from the George Reserve, Michigan.* In L. K. Halls [editor] *White-tailed Deer: Ecology and Management.* Wildlife Management Institute, Washington, D.C., and Stackpole Books, Harrisburg, Pennsylvania. pp. 211-242.
9. O'Bryan, M. J., and D. R. McCullough. 1985. *Survival of black-tailed deer following relocation in California.* Journal of Wildlife Management 49: 115-119.
10. Ricker, W. E. 1954. *Effects of compensatory mortality upon population abundance.* Journal of Wildlife Management 18: 45-51.
11. Savidge, I. R., and J. S. Ziesenis. 1980. *Sustained yield management.* In S. D. Schemnitz [editor] *Wildlife Techniques Manual.* The Wildlife Society, Washington, D. C. pp. 405-409.
12. Scott, R. F. 1954. *Population growth and game management.* Transactions of the North American Wildlife Conference 19: 480-503.
13. Wagner, F. H. 1969. *Ecosystem concepts in fish and game management.* In G. W. Van Dyne [editor] *The Ecosystem Concept in Natural Resource Management.* Academic Press, New York, New York. pp. 259-307.

8

Aldo Leopold
and the Real World

Huey D. Johnson

In the real world, resource allocation increasingly ensures conflict. Money needed for managing the quality of natural resources is constantly shunted to competing purposes, like Star Wars. All the while concrete is poured for irrational projects, water is polluted, soil is lost. Such conflicts can be dealt with and enough victories won to give us hope for the future, but only if there is a human force to balance the simple economic arguments given to justify any development. Skilled advocacy in the public arena, backed by information from specialists, will advance environmental progress. I decided to take that career path years ago.

Aldo Leopold was important to my decision. Moreover, his writings and personal example have had an enormous influence on this nation, an impact that is yet little appreciated, but one that will endure. I believe Leopold will become as well known as Thoreau in the next century.

A Personal Perspective

Although much has been written and more needs to be written about Leopold's land ethic, his aesthetic, and other ideas, there seems to be little recognition of his substantial role as a leader-activist. More than any other source, it has been the ideas from Leopold's writing and the example of his activism — to carry out action in addition to thought — that inspired my career.

I first read *A Sand County Almanac* when I was snowed in at Lake Tahoe on a fisheries research project. The night I finished it I wrote to my wife-to-be that I had found my drumbeat. Over the years I continued the march with some satisfying results.

As an advocate of *A Sand County Almanac*, I have given away more than 100 copies of the book. I suppose at times it has been fun to thrust a book at some other overzealous advocate before that person could give me his religious tract.

There have been some nice ripples from this activity too. Amid Leopold's admirable stances was his willingness to question a headlong plunge into a technology that at first was believed to be the solution to all of society's needs. Remember his comment about the "great god Ford" and its impact on the quality of life as well as wilderness? What may be the best result of my book distribution effort was the inclusion of some of Leopold's ideas in what has become the most popular general introductory college textbook on physics (*Conceptual Physics* by Paul Hewitt). It happened after the author and I exchanged books and then ideas on the need to balance technology and environmental quality.

Hewitt and other scientists have expanded the budding scientists' worldview of science-technology and nature. We resource professionals need to catch up and get the future crop of resource managers involved. The next wave of policy development needs to include involvement from our side. Genetic experiments, new substances from a mixture of microbes and technology, the improvement of management practices that comes with computers, telecommunication and space photography are just a beginning. We cannot afford to keep missing the boat as we did with nuclear development. A good start would be to make *A Sand County Almanac* required reading, for the philosophy and ethics expressed therein apply well to the future.

Thankfully, there is out there in the real world a thin green line of people who care — call them resource professionals or environmental activists. They are the unsung, committed people who are responsible for and carry out to the best of their abilities the ethical allocation and management of natural resources and environmental quality.

The scale of their professional responsibility is awesome, and the setting in which they work is difficult. Awesome because, as Theodore Roosevelt stated back in 1908, "The prosperity of our people depends directly on the energy and the intelligence with which our natural resources are used. It is equally clear that these resources are the final basis of the national power and perpetuity." No president since Teddy has understood that or cared enough to learn from history. This is especially true today.

History offers many examples where concern for the vitality of the land was ignored, and people — whole nations — suffered. The fall of the Roman Empire was predicated by half a dozen reasons, not the least of which involved the desertification of the soil due to poor irrigation practices. In this respect, like North Africa and other arid parts of the world, my home state of California is on a similar decline, with 300,000 acres of prime agricultural land already out of production and much more affected.

There are those who would say that is changing. To respond I need only turn to the Middle East, with its ongoing conflicts over oil extraction, desertification, and water. Last year, Egypt's minister of state for foreign affairs was quoted as saying, "The next war in our region will be over the waters of the Nile, not politics."

In responding to such problems in the real world, Leopold's ideas are important to those seeking a path to a better future.

Leopold's Impact on Resource Professionals

Many of us who are resource managers or conservation activists would be proud to be known as Leopold people. Most previous resource management thinking had a major economic emphasis based on Gifford Pinchot's beliefs. Leopold went beyond that. He put resource management into a workable context.

The most eloquent statement I have found in his writing on the subject said: "The objective of resource management was to create a more enduring civilization. Management involved the use of tools — economic, legal, political, scientific and technical, not just one of them." Leopold's writing defined ecology — how the land and living things maintain their existence — and its relationships. Not just a tree or a forest, but interrelationships with soil, wildlife, water, beauty, and recreation.

By following Leopold's lead, rather than Pinchot's, we could function realistically, daring to consider applied approaches; our minds could be free. A scientific base, but not limited to scientific emphasis. Management, but not limited to that. Leopold's writings described the beauty of the natural world and its constraints. By being free from strictures of specialization, he was able to synthesize other factors, such as economics, history, and religion.

To me, Leopold's approach made him a prophet of possibilities, and there are some themes in his life and work that I view as relevant for making things happen in today's real world. They all revolve around his concept of resource management, which includes what I call environmental systems management. I will give you examples of his themes through my own experiences. But more important, I have chosen five additional areas that show how Leopold has been a mentor to me and my actions on behalf of the environment. These are the role of intuition, on being a generalist, inspiration, leadership, and Leopold's role as a hero.

I feel strongly that systems management is a method we need in order to attain the quality and productivity of our national environment. Much of Leopold's thinking and actions opened up an exciting, new direction for a systems approach to management. That was a lesson I had learned in an earlier corporate existence. In the real world of decision-making one soon learns that few things are simple. There are usually tangible and intangible factors in abundance, and one must deal with the relationships between them. The problem is that there is not time to sort thoroughly through them all. You decide, then act on the accumulation of information and an extra sense that has no tangible definition and move on.

Intuition. One learns to be comfortable with fundamental insight as an internal guide. Needed knowledge, experience, and ethics that are part of it can be gained through a training process, in an institution or alone. But it is not something that is seemingly "taught" in college today. If a university football coach can condition players to function well alone in the stress of a real game by sharing his experience, we need to do the same in training resource professionals.

Resource students may not have to face these conflicts on Saturdays while they are in college, but the stress arrives quickly as they get into careers that involve resource allocation. Because we send them out without the training necessary to handle those conflicts, national resource policy suffers. One outcome is that when they attain positions of power they are not astute enough to hold onto operating funds to run agencies. This leads to fewer future jobs for stuents who really are needed to help bring back the natural productivity of America.

Leopold understood the need to include this internal guide — intuition—in all decision-making, as is obvious in this Leopold quotation from Susan Flader's biography, *Thinking Like a Mountain,* "Possibly, in our intuitive perceptions, which may be truer than our science and less impeded by worlds than our philosophies, we realize the indivisibilities of the earth—its soil, mountains, rivers, forests, climate, plants, and animals, and respect it collectively not only as a useful servant but as a living being."

Though he rarely wrote about the philosophical origins of his ideas, this one is traceable in an interesting way. Leopold had a well-used book by a Russian philosopher, P. D. Ouspensky, who "regarded the whole earth and the smallest particle thereof as a living being, possessed of soul or consciousness."

A Generalist. Aldo Leopold was not perceived as a dilettante because he was informed enough to be respected by specialists while not confining himself to a speciality. The breadth of Leopold's thinking gave many of us courage to think that being a generalist was possible.

An advantage the generalist has is the ability to carry an idea from specialist to specialist, getting their assistance in developing the central idea, but holding on to the decision-making. The opportunity for synthesis thus lies most with the generalist who progresses from each information realm.

Political involvement is essentially a generalist's responsibility. It may be *the* most important role in management. All actions between groups of people involve some form of politics. Getting budgets—money for institutional operations—includes a political process. At present, the lack of generalist political involvement is a prime reason for the disastrous budget trends in resource management agencies.

Inspiration. Leopold's writings are my philosophical touchstone, carried always in my mind as a source to which I can relate. His writing was of such quality that it projected information with a sense of beauty, of the greatness and grandeur of nature and evolution, of the interplay between humanity and natural ecology, and of the obligation to manage for permanence, which could only be achieved by a functioning land ethic.

Leadership. Leopold thought, then he acted. If necessary, he took risks in standing up for his principles. That represented the most valuable quality of human affairs—leadership.

The final measure of the quality of leaders' lives is courage. Leopold, in my mind, stands beside several others who had it, including Einstein and Churchill. There are ample examples of Leopold's courage in his new biography by Curt Meine.

The time is coming, and we as resource managers must help it arrive, when there will be increasing priority on improving resource management as public policy. For many involved today professionally, there will be an opportunity to lead.

Leopold as Hero. To inspire the majority a few must risk taking an activist role. Leopold's willingness to take action beyond his written word was certainly inspirational, but it is the action itself that makes him a hero to someone. After Leopold had organized the sportsmen of New Mexico, he then took the group's leaders to call on the state's newly elected governor. The governor said he could not go along with what Leopold wanted, whereupon Leopold remarked, "In that case, Governor, the hunters and fishermen of this state will not vote for your re-election." He then got up and left.

Students in most fields — physics, music, law, religion, art, athletics — have endless numbers of heroes against whom to measure themselves. Many a student, for instance, gains extra strength by thinking of Einstein's dogged persistence in his research.

The same occurs professionally for a natural resource manager. There is a need to have a hero to think about, primarily because you must accept conflict and at times stand alone as part of an ethical resource management role.

For me the help of a hero has been called upon at the loneliest moments, on the margins of uncertainty. It was then that Leopold's principles provided comfort. One needs that to keep going in the roughest of times.

The Wild and the Urban: Two Examples

As California's secretary of resources, I had the opportunity to change policy in various ways. I planned to solve a number of issues that were always points of conflict. That would allow future resource management to proceed more effectively. Getting federal and state wilderness lands designated meant we could preserve wilderness and also get the best public tree-growing lands defined and into productive management. Wild rivers was another of the complex, conflict-ridden issues with which we dealt.

The wilderness issue had been a raging conflict for years in the western United States, especially as it concerned national forest land. Wilderness was an idea that demonstrated Leopold's courage. He defined it, called for it, and forced the establishment of the first wilderness area—the Gila.

In my case there was much give and take as several sides attempted to enlarge or reduce any wilderness on Forest Service land in California. There were 6 million acres involved. The nonprofit groups had pushed wilderness designation for years, and over those years I had been part of their effort outside of government.

But things finally came into focus. The Forest Service presented me with a ridiculous proposal. It involved the wrong locations and far too small an acreage. I rejected it, then went directly to the forest products industry; we started successfully negotiating individual areas, agreeing on prime, productive areas that did not need to be wilderness and wild, rugged regions that did. The Forest Service decided to block that effort and approached industry separately, offering industry whatever it wanted. Industry negotiators then came to me and simply said, "Look, there is no logic in our negotiating further. They have given us more than we ever dreamed of getting."

From there, instead of negotiating further with the Forest Service, I went directly to Congress. California Representative Phil Burton agreed to champion the issue.

In Sacramento, and in Washington for that matter, intense pressures opposing wilderness came from every direction. The pressure built, anger increased. The governor was besieged with appeals on both sides.

Things bogged down. The side that opposed wilderness was gaining a position to block the legislation by offering feeble compromises. My response was to threaten a lawsuit against the Forest Service.

Those on my side in the private sector—the Sierra Club, Wilderness Society, and Natural Resources Defense Council—my staunchest friends and allies—came to my office and said, "Please don't file that suit. You will rock the boat and destroy the compromise that we are working out." I told them that they were giving away the issue by compromising when they need not compromise.

When they left, I was as lonesome as I had ever been. My friends had deserted me on the issue. I faced the morning knowing that when the governor learned of their opposition I would not be able to file the suit. After reading a bit of *A Sand County Almanac*, I went for a walk, then called the attorney who had the papers ready at his home and asked him to file them.

I won, fortunately. The suit was challenged by the Forest Service at a higher level and the state again won. That stopped the Forest Service and the compromisers dead in their tracks. Millions of acres of wilderness were achieved that would not have been otherwise.

Remember the line in *A Sand County Almanac,* that "weeds in a city lot convey the same lesson as the redwoods.... Perception, in short, cannot be purchased with either degrees or dollars"? It is an apparently innocent statement, but over the years it has been basic to my hopes that urban dwellers would become enlightened by and involved in environmental issues.

I was involved in helping to save redwood forests and lots of other unique landscapes while working for The Nature Conservancy — and actually acquired the tallest redwoods for preservation. This was not a major accomplishment compared with the magnificent effort by Save the Redwoods League, but I offer it as a sample of personal involvement in the redwoods issue.

Then came reapportionment: one person, one vote. The advantage in Congress of rural interests was no more. The cities would, because they had the numbers, dominate things as time went on, including resource policies for the nation.

As one interested in resource and environmental policy, I first worried about the environmentalists' major emphasis on the wildlands found away from the cities. To save wildlands we had to have the city folk as allies in future policy struggles.

The weeds-in-a-city-lot theme mirrored my concern. We did not need to bring city dwellers to the wild, preserved wilderness; we could use the opportunities in the city. Then those who chose to do so, those who were perceptive, would appreciate and use the public lands that were their birthright.

In a practical way, how could we affect urban awareness? As an individual, I could only proceed on the basis of experience, and that was limited to a certain expertise gained while acquiring land with rare life forms for The Nature Conservancy. Leopold's weed line flowed through my mind, and I decided that the only practical result I could get would be to try and save lands of relevance to people in the cities. The form it took became the outline for the Trust for Public Land.

I founded the Trust and ran the organization for its first five years. It has been very successful. By 1985 it had saved more than 300,000 acres of precious landscape to be passed on to public agencies at a savings exceeding $30 million. There are endless, wonderful stories to share: from bums panhandling money to save their own garden parcel in the Bowery to high-risk preacquisition of thousands of acres for a proposed national recreation area adjacent to the Golden Gate Bridge. This area has become the most used of any federal park in the nation. To get it we had to block a new town development on land owned by Gulf Oil and acquire three miles of coastline owned by RCA. Both companies, though initially upset, were pleased with the final result. Both their image and their bottom line financially worked out over time.

An Experimental Measure

Impacts must be measured by the effects on others. To test a hunch on the matter, I conducted a poll on Leopold's impact on resource professionals.

It was a year ago that the annual meeting of the Association of Western U.S. Fish and Game Commissioners was held in Colorado. The association's name is misleading in that the meeting draws top management professionals regardless of the area of their responsibility: foresters, water managers, range managers, wildlife managers, educators. It is a meeting where much happens in the halls. Politics and pressures are discussed and compared; people discreetly look for top-level jobs, and so forth.

I asked everyone I was about to address to complete my questionnaire. It asked all participants to list the three sources of information most important to their professional career. The response was marvelous: Shakespeare, Dasmann, Abbey, and others. But 90 percent of the people listed Aldo Leopold, and, interestingly enough, not one mentioned Gifford Pinchot. A majority of those professionals have top-level jobs managing the public land resource.

To this point I have said little about the impact on the environmental ranks in the nonprofit or activist sector. Without fail, I have never talked with anyone in the Sierra Club, the Audubon Society, the National Wildlife Federation, and other groups who did not know and respect Leopold's work.

I mentioned wilderness. I am proud to be on the managing council of the Wilderness Society, an advocate group that has worked

hard for wilderness. Leopold, Bob Marshall, and others thought up
this organization while driving down a road in the Smoky Mountains. They simply got excited at the prospect, pulled over, and put
the idea together.

The Future

The future is now; it is time for action. History beckons us in this
year of the Leopold centennial, for resource-related things have
taken a bad turn. Not only are resources affected directly, but the
resource management agencies responsible for the stewardship of
the nation's base of vitality, as Theodore Roosevelt described it, are
being devastated, and there is a real danger that the problem will
become worse.

One of the problems is the Reagan Administration, whose level of
environmental knowledge belongs in an earlier era, when the nation
believed there were no limits: the nation and world could house and
feed any number of people, soils would last forever, the waters
would flow unpolluted and remain abundant forever, the air would
remain pure, and there would always be another virgin forest over
the next mountain to be logged. President Reagan is a throwback to
the first 100 years of the nation's history when exploitation was the
practice of the day. Under his guidance, the quality of soil, water,
air, forests, parks and wilderness, prairies, wildlife, plants, and
climate, which needs to be managed and maintained for what
Leopold called "an enduring civilization," is being sacrificed.

The seriousness can be seen by taking a close look at resource
management agencies. Though any number of now battered agencies could be selected, let's look at Leopold's first employer, the
Forest Service, which is responsible for managing millions of acres of
the nation's forest, watershed, grazing, recreation, and heritage
lands and on which every American can find peace and beauty.

The Forest Service, once a proud, effective agency, is in a sad
state philosophically and operationally. In the process of conceding
the quality of its lands to exploitation, the Forest Service has had its
operating budget cut 20 percent over the last six years. It has money
to build roads in the wilderness, but few funds for the recreational
development needs of the public.

To me the agency's worst error is its lack of attention to soil maintenance. Not only the future of forest crops but the future health of

the nation hinges on soil quality, and the Forest Service has not maintained soil as a priority issue.

Friends of the Forest Service become fewer by the month. The only cheers I have heard for the agency in a long time come from the handful of companies that enjoy the gifts of public timber. The Forest Service loses money in the process of timber harvesting and thus required a subsidy of $600 million from the U.S. treasury last year.

The agency's notoriety was finally acknowledged by the White House. The Forest Service suffered a slice by the backswing of Reagan's budget-cutting sword when the president criticized the agency in his most important speech of the year to Congress and the American people, his State of the Nation address on the budget. In that speech he named some federal programs that overextended, misdirected, or operated on too expansive a scale in the current tight budgetary environment. On the list were "postal subsidies, interstate highway grants, the Forest Service, and many other programs."

In the fall of 1986 the Forest Service announced it would double logging on the national forests by the year 2030. This is something of an impossibility considering the lack of productive management funds to deliver a future forest harvest and the fact that most of the logical and thus profitable landscapes to log are already cut.

Forest Service officials currently defending these practices of giving away the public resources should be wearing striped suits in prison rather than Forest Service green. Their policies have as much logic as allowing bank robberies so long as the money stolen is spent in the county where the robbery occurred.

Aldo Leopold was proud to be part of the Forest Service. But it has changed. Leopold is clearly the most famous of Forest Service people, but he has not been part of the agency tradition to date. It is still dominated by the narrow economic focus of its beginnings. Gifford Pinchot, its founder, was a magnificent pioneer, who deserves much recognition. His professionalism lives on as an example for us all. Unfortunately, his emphasis on economic forestry is badly outmoded today.

Instead of understanding Leopold as a dominant guide for enlightenment to resource management, one that I believe will hold a permanent place in the minds of people, the agency's leaders have ignored him, or worse.

The Escudilla Mountain is a case in point. Remember that chapter in *A Sand County Almanac* about the great mountain of "think-

ing like a mountain" and the story about the death of the last grizzly
in Arizona? The Forest Service until recently had 25,000 acres there
that were roadless. It could have saved that mountain's wilderness
area. That would have been a suitable shrine to Leopold. Instead it
butchered the area with roads and logging so that the wilderness remaining is the minimum required by law—5,000 acres. Even the
mountain sides were not saved.

All this is to say the Forest Service has lost its proud step and is
staggering toward its death to a dirge with muffled drums.

There is hope. I predict the American people are not willing to
allow an elected official to ruin the nation's natural wealth. The
public has grown far more aware than Smoky the Bear. Poll after
poll shows that 80 percent of the American people place environmental quality as a national priority, ahead of economic and other
factors. The latest poll I have seen was a Canadian study done in the
summer of 1986 in which the Canadian public placed the quality of
the environment ahead of other factors, including jobs or other economic concerns, by a margin of 93 percent.

The time to reverse that thinking is upon us. What would Leopold have done? Acted, I think. We must do the same. We must act
to reverse the disastrous onslaught the administration is imposing on
our heritage.

As one goal, we must define how we can assist the resource-man-
aging agencies to recover, to get their share of the treasury so they
can operate and deliver a productive future. That includes saving
the Forest Service. The agency has fine managers and young professionals coming up. The need is to provide some legislation so the national forests can be managed with integrity for the long term.

Our opponent is vulnerable, and to be successful in the real world
one must know one's opponent. There is a classic opportunity to do
that. Read David Stockman's book, *The Triumph of Politics*. A former director of the powerful Office of Management and Budget, he
tells all and describes especially well the weaknesses of the president.
Rarely has this happened while a president was still in office.

In a recent magazine interview, Stockman confirmed a key point:
The president will not stand with a budget cut on an item when
there was a demonstration against it by the public.

I offer this strategy in a professional way—and with discomfort
because one does not like tackling the president. It is difficult to do,
first of all, and he *is* overworked. Moreover, he is an honorable man,

I am sure. But we, as Leopold people, have an obligation to proceed with a mailed fist if a gloved hand does not work.

The president's attempt to exploit the public resources has historical precedent. Napoleon's fiscal crunch from war led to his selling to our nation the Louisiana territory for a pittance, and the Russian czar who needed money for his overblown military expenditure sold us Alaska at a bargain price. Should we fear that Mr. Reagan will follow in their footsteps by offering to sell Alaska to Japan? It may be the only honest way of paying for Star Wars and other fantasies.

All restrictions seem to have been lifted from any logic of financial management. Themes like Star Wars and billion-dollar subsidies for a failed nuclear energy technology are dangerous precedents. In a feeble, desperate attempt to pay for these fantasies, our president has sacrificed our nation's public resources and resource management agencies as well as the universities that train our professionals and generations of future Americans.

I would look forward to commending the president if we could get him to change his mind, to understand why our position is a correct one. He is not entirely to blame. There has been no advocacy of the themes that the professionals managing our resources could provide. We have allowed narrow specialities to retreat to caves where they can be isolated and beaten down.

There is a strategy of mutual assistance. It fits a model provided by our founding fathers. Resource agencies can work toward a common goal once their leaders accept the reality that they must do so. To borrow a phrase from the Constitution era, we hang together or we hang separately.

To commemorate Leopold's 100th year, I propose a crusade. Our goal: to manage resources for permanence, or what Leopold called a more enduring civilization. We need to create enough public support to guarantee a vigorous, ethically managed, productive American landscape as a permanent American heritage.

The new coalition behind this crusade, to be called the Strong Oak Movement, needs to be broader than just environmental groups so it will be exciting enough to be heard and respected in the public arena, as we seek an upgrading in policy support for improved environmental management of the public trust by government.

With that in mind, I have a wish: That there were some university or training institution that had as its goal the preparation of young leaders who might become resource professionals like Aldo Leopold.

Estella and Aldo at the Shack (circa 1945)

9

The Land Ethic: A Guide for the World

Bruce Babbitt

Aldo Leopold belongs to Arizona, my home, in much the same way as he belongs to people in Iowa, Wisconsin, and the tall-grass prairies of the Middle West. What I understand best about his legacy, I understand from having stood where he stood and having looked out at the wilderness he saw. Growing up, as I did, on the rim of the Grand Canyon and spending my summers hiking the back country, learning its geology, its history, its ecology, I learned to appreciate what Aldo Leopold championed for all Americans.

Leopold was born and raised in Iowa. He spent his youth exploring the marshlands and fields around the Mississippi River near his home, taking in the air of the great prairie, the water of the river, the hardwood forests, and the endless fields of grain.

As a young man he was sent by the Forest Service to Arizona to map timber stands in the White Mountains.

Seventy-five years later I took my own son into the White Mountains. We stood and wondered at what Leopold described as "a great meadow, half a day's ride across.... The edges of that meadow were scrolled, curled and crenulated with an infinity of bays and coves.... One often had the feeling, riding into some flower spangled cove, that if anyone had ever been here before, he must of necessity have sung a song or written a poem."

In my son's eyes and in the questions he asked, I could see that, standing there, he was beginning to appreciate what Leopold would have had all Americans understand. I remember, in particular, my son asking a child's question, "Dad, are there wolves living here?" I went home and read to him from *A Sand County Almanac* the description Leopold gave of the wolf he had shot—how he had looked into her eyes and saw the green fire dying and understood that in that fire was something only the wolf and the mountain had understood until then, something he now understood as well.

The green fire Leopold left in us was the land ethic, an idea that humans are citizens of nature, members of a global biotic community and must begin to think as such. He wrote about "a state of harmony between man and nature." "A thing is right," Leopold said, "when it tends to preserve the integrity, stability, and beauty of the biotic community. It is wrong when it tends otherwise.... We abuse the land because we regard it as a commodity belonging to us. When we see land as a community to which we belong, we may begin to use it with love and respect."

In his own time Leopold's words were the impetus for the conservation movement, for the development of the Forest Service. He was the first to suggest a national wilderness system, and in 1964 that idea became the National Wilderness Act. His work in preservation of wilderness goes on in our continuing fight to secure the unfinished American wild, to gain what spiritual enrichment we can from that wild, and in our concern over preservation of the wilderness around the world. The destruction of wilderness, along with deforestation and species extinction, continues to plague this globe and demand our attention.

A Moral Counterforce

In our time the land ethic is still the crucial moral counterforce in opposition to short-term expediency in determining the use of our land, air, and water resources. It is still a warning that every person must see himself as a citizen of nature, actively living out the goal of preserving the integrity and beauty and stability of the natural world.

That lesson and that warning are timely indeed. The policies of James Watt, Anne Gorsuch Burford, and the Reagan Administration have come as a rude reminder that decades of care and stewardship can be erased in no time and without the slightest difficulty by those who do not understand. At the same time, it is not just our government but our own increasing power to wreak profound havoc on the biosphere—havoc that knows no regional boundary. It is that power that calls us to rethink clearly our commitment to Leopold's land ethic.

How does that land ethic, which addressed principally public land in Leopold's time, apply now to private land and to the international arena? The most interesting thinking on environmental issues today is not going on at the national or the state level. It is international in scope, and it is local in scope. It is thinking that, on the one hand, is as small as our backyard and, on the other hand, as big as the globe.

The first issue on the agenda must be a global understanding, one that does not require periodic international disasters to remain in our consciousness, an understanding that we have magnified our power over the biosphere and that we have the potential to destroy it. What we do wrong in one country today is left as punishment in another.

Earlier this year the Soviets suffered a "nuclear mishap," one they kept from the rest of the world for a number of days, one whose importance they still try to minimize. What we know now about Chernobyl is that the Soviets had an atomic fire burning on their soil. It was a fire that will leave that region of the Ukraine uninhabitable for generations, a fire that will claim, in time, more than 20,000 lives. It is a fire the fallout from which, in the end, will equal the fallout from every nuclear explosion ever detonated, from 1945 to 1986.

Chernobyl was not a regional problem; it was not a national disaster. It was instead a prototype of an international catastrophe that we have only recently been able to create for ourselves. It is a reminder that environmental protection is an international affair, a reminder of nature's most basic rule—that everything is connected to everything else.

Some fallout from Chernobyl fell on a part of Sweden inhabited by the Laplanders. These Laplanders depend for their survival upon the reindeer they herd. This year, however, the Laplanders' reindeer are unfit for consumption or for sale; they have been contaminated with radioactive fallout. Because of Chernobyl, a nuclear accident not far from Kiev in the Soviet Ukraine, a culture and a way of life that have gone on for centuries in the cold North may end for good in the year 1986. That is because the human animal is part of a food chain that depends on the well-being of animals below it and on the earth itself. Forgetting that, humankind can hurt itself indirectly in an infinity of ways.

A Matter for International Negotiation

What the world needs to do, not just the United States, is to come together to address problems that we have no business thinking of only as national issues.

It may seem strange to suggest that leaders of the world should meet to deal with issues of the environment. But that is just what I am suggesting. Environmental issues ought to be on the agenda whenever Ronald Reagan meets with Mikhail Gorbachev. The extraordinary nature of man's ability to poison his own environment demands extraordinary attention.

In 1908 Teddy Roosevelt called all the governors together to confront issues of natural resource conservation. Those governors had never met as a group before. It was an extraordinary gesture. In 1986 we need an international analog to what Teddy Roosevelt accomplished. We must say to the rest of the world what Teddy Roosevelt said to the governors that year, "We cannot...when [we]... become fully civilized and very rich, continue to be civilized and rich, unless the [world] shows more foresight than we are showing at this moment as a [planet]...."

As a country we have to be as anxious to bring to the rest of the world our national respect for the environment as we are to export our belief in democracy and a market economy. These three together should be our legacy.

That international approach is necessary not only to deal with issues of nuclear power but to deal with the devastating effects of modern industrialization. Let me outline just a couple problems that merit international attention.

The first of these is the possibility of climatic changes due to the increase of greenhouse gases in the earth's atmosphere. The greenhouse effect is caused principally by the use of fossil fuels that release carbon dioxide and chlorofluorocarbons into the atmosphere, a release that threatens global warming. Atmospheric warming could cause the oceans to rise, reduce croplands, and extend the deserts to new regions of the globe. All of this is creating massive disruptions in the global economy and demanding profound changes in our use of fossil fuels.

The threat requires action, not just study. We should place the issue on our negotiation agenda with the Soviet Union and begin a worldwide dialogue on the problem. The truth is we must rededicate ourselves to confronting the problem of world energy management because the recent slip in petroleum prices and discord within OPEC have threatened to erase all the progress we made in energy conservation during the 1970s. We cannot allow ourselves to forget everything we have learned within one short decade.

A second problem of similar international scope is that of acid rain. There is no point in repeating what has been said on this issue. We understand the effects of acid rain on our forests and our lakes. We have heard the complaints from such countries as Canada and Norway. We have also heard the silence of the United States and Great Britain. We know the policy of the Reagan Administration — ignore the fish and the forests and maybe they'll go away. Instead, I want to tell you an acid rain story, a cautionary tale with an ironic twist from my own State of Arizona.

Arizona is famous for copper smelting. The industry employs a lot of our citizens. We were confronted by an application from the Phelps Dodge Company to continue to operate a smelter at Douglas — a smelter that was emitting vast amounts of sulfur and clearly violating environmental laws. We finally went to court and closed it down, only to learn that right across the border in Mexico another company was building a smelter that would double the sulfur emissions in the western United States. So we Arizonans know how the people of Canada and Norway feel. This is not an indictment of any one government, but an appeal to stop talking, stop studying, and start acting as one world to solve the problem. At the same time, in this country we have to enact strong measures at the federal level. I stand with 170 congressmen and women who endorse H.R. 4567, the Sikovski-Waxman Bill. The bill is fair to every region of the country and can get us to work quickly on the acid rain problem.

All of this is to say that we must take Leopold's land ethic — that sense of humankind taking their places in the biotic community — and apply it within the world community. That is the logical step.

A New Wave

The other arena in which the land ethic must take root, and where I think it is in fact taking root, is in our own backyard. Aldo

Leopold's other great contribution was the idea of a personal conservation ethic. I think we are seeing just that. I think we will look back on the 1980s as the decade in which the environmental movement went home—left Washington to find its grassroots support, to relate to people, to sharpen its message, and to gain energy for a new wave of reform. In the 1980s, as Washington slept, there grew public support for public lands and public awareness of public health. Today, community study groups come together to fight for wilderness and against toxic pollution. In this decade the membership of the Sierra Club has nearly doubled. That is the conservation ethic at home.

I see it working in Arizona where community groups and environmentalists joined together with government to save the San Pedro River and the wildlife, like the black hawks, that live there. You can see it working in the Northeast, where homeowners and small towns push and push to keep toxics from being dumped in their backyards or to have them cleaned up when they have already been dumped. You can see it in public awareness of the radon gas problem.

You can see the conservation ethic work in Iowa also, as people in the state begin to realize the price they will pay for the continuous spraying of insecticides and herbicides and fungicides on their fields, poisons that seep into the ground and infect the water supply and milk supply. Anyone who has spent time in Iowa knows that its citizens are alert to the problem and have begun to take action with government to solve it.

Iowa, like Arizona, like New York and New Jersey, has experienced a renaissance of the personal conservation ethic. It is a renaissance tied on the one hand to a selfish desire to protect our quality of life, to clean up our own backyards. On the other hand, it is tied to an unselfish respect for the natural community that Americans have always had since they were pioneers.

It is a respect that must be taught to our children—appreciation for the goldfinch and the wild rose and the oak and the tall-grass prairie. Simply put, it is the respect that led Aldo Leopold to die on his own land—protecting that land and his neighbor's land from fire. His death is a symbol of what Leopold, John Muir, and René Dubos might have said to each of us—the time has come when we must "think globally and act locally."

Leopold would have expected that of us—out of respect for the land—and as a sort of proof that we have love for the children who will follow us.

III

From Burlington to Baraboo

Aldo at Les Cheneaux Club

10

Aldo's School Years: Summer Vacation

Frederic Leopold

Our summer routines were tremendously affected by Mother's severe hay fever. Seeking relief for her, our grandparents discovered a unique summer resort, the Les Cheneaux Club, located about 15 miles east of Mackinac Island along the north shore of Lake Huron. The area was located in a small remnant of comparative wilderness, on the edge of the Canadian Shield; the ground was just able to yield a stunted growth of north woods trees in the thin layer of soil barely covering the solid bedrock.

There were no roads or railroads and little exposure to the populated Midwest. Access was by water only. The club and its members owned a point of land about a mile in length, extending from the northeast end of Marquette Island, one of a group of small islands clustered along the north shore of Lake Huron. The small water between the islands offered protection from the waves of the big Lake Huron. The tiny villages of Hessel and Cedarville on the Michigan mainland nearby were populated by the only year-round residents within 15 or 20 miles. Our cottage supplies of food and miscellaneous items were delivered to us daily by launch from their village stores. The term motor boat had not been coined at this time.

Our own cottage was small and simple, overlooking a bay nearly a mile wide to the west. Summer sunsets over the bay were often indescribably beautiful. It was here too that we enjoyed swimming and water sports in general. No wonder that we children during our home months dreamed of our summer pleasures and actually count-

145

ed the months, weeks, and days until we journeyed to our summer cottage.

The Journey

Usually, we departed for the trip in late July or early August and we remained until about September 26 when hay fever ended for the season. The weeks before starting north were filled with preparation and anticipation. We all offered to help Mother with packing the many things we would need. As I recall, our heavy baggage filled at least six trunks, plus a bedding roll for our sleeping bags; a special trunk contained camping gear of all kinds. It was the duty of we boys to pack this trunk.

The local drayman arrived the evening before we left to haul the trunks to the railway station. He was a powerful man whose broad back enabled him to carry even our largest trunks down our winding stairway without assistance and with never a scratch on the wall enroute.

Baggage was delivered and checked at the railroad station the day we left. Our train left Burlington at 4:30 a.m., arriving in Chicago about 9:30 a.m. This left us about two hours to reach our boat dock on the Chicago River from which the palatial steamship Manitou departed about 11:30 a.m. The transfer across town was made in a horsedrawn Parmalee Coach from Union Station to the dock. These coaches were usually crowded with passengers and hand baggage was piled on top. There was no room for our dogs, so my older brothers were entrusted with walking the dogs across town on a leash.

The big ship was most exciting, and we felt privileged to be able to enjoy the 24-hour voyage to Mackinac Island. Much of the way we followed a course up the middle of Lake Michigan, so land was often out of sight.

None of us was affected by seasickness, except Aldo, and he hated this evidence of physical weakness. We carried dried dog biscuits for our animals, and Aldo found he could eat dog biscuits without getting sick.

In the morning, before reaching Mackinac Island, the Manitou made stops at Charlevois and Harbor Springs, where many summer residents came to the dock to observe the big ship. Mackinac Island was reached about noon, leaving us with a two-hour wait until the

S.S. Islander departed on the last lap of the journey. We all enjoyed the beauty of Mackinac's white cedars and the whitewashed stone walls of the old fort overlooking the Straits of Mackinac. The main street was lined with gift shops, usually crowded with tourists buying souvenirs of dubious quality. No automobiles were allowed on the island, and fancy surreys with fringed canopies, drawn by well groomed horses, offered short or long rides to various island points of interest, such as the fort, Lover's Leap, and the British Landing across the island.

Eventually, we boarded the old familiar Islander with Captain Mac Carty in command. We had been coming for so many years that he recognized us as old acquaintances.

The last leg of the journey covered about 15 miles and took about an hour and a half. We watched for landmarks, such as Goose Island, Point Broulie, and familiar cottages, then the first stop at Hessel, Michigan, where Fenlon's store provided daily grocery deliveries to our cottage dock at the club.

Eventually, we rounded the last point and came into view of our club. Summer guests at the club usually met the boat, and among them were old friends. Lots of college sweaters with "Y" for Yale or "H" for Harvard or "P" for Princeton were in evidence. It was a great place for young people as well as their parents, and especially for infants accompanied by nursemaids.

Our cottage was opened by the resident steward of the club prior to our arrival. That meant the board shutters were removed, the plumbing turned on, the coal oil lamps filled, stove wood and fireplace wood provided, and rowboats launched. In a very short time life at the cottage was in full swing.

A Land of Water

Aldo did not care for golf or tennis. He preferred exploring the woods, sailing, and fishing for smallmouth bass and northern pike. Marquette Island was of irregular shape, with long points extending from the main body. Overall, it was five or six miles long, north and south, and about four miles east to west. Aldo knew most of the island intimately. He produced several handmade maps artistically decorated and illustrated with typical trees, animals, and birds in appropriate places. All of the trails were shown, including some newer trails that he himself created. Almost all local travel was by water; there were no roads. One of the servants at a neighbor's cottage once remarked, "What a land of water."

Les Cheneaux Club was centered around a large-frame, two-story clubhouse with guest rooms available, and the dining room served meals to cottagers who preferred not to run their own kitchens. There was a small dance floor where simple dance music was played on two or three instruments, six evenings a week. Weeknight dances were simple and ended by 10:00, but all the young people learned ballroom dancing. Aldo was an excellent dancer. (Later in Madison he and his wife Estella were picked as the best dancers at a faculty party.)

The club property was located in an old burned-over area. The new-growth birch, aspen, spruce, and balsam were no higher than the eaves of the cottages. For the benefit of nursemaids with parambulators and older people who wanted mild exercise, the club had constructed a boardwalk that ran past the cottage paths and on through a woodsy area. It was laid out in a long ellipse, so the overall length was about one mile, whence its name—the Mile Walk.

Once on the walk, Aldo encountered a wandering skunk, or, as Father called them, a "sachet kitten." Aldo killed the skunk and proceeded to immortalize its memory by carving into the crossboards of the Mile Walk his declaration, "Aldo Leopold killed a skunk here on August 20, 1901," for all to see. Years later the boardwalk was replaced by cement, but the carved boards were salvaged and nailed in their proper order to two birch trees nearby.

Among Aldo's discoveries on the neighboring mainland was an old bear trap. The trap was constructed of logs arranged to drop a heavy burden of logs on the bear's back if the animal tried to extract the smelly bait. He discovered rabbit traps as well, also of the deadfall type. Lines of these traps, located along the rabbits' travel paths, were operated in winter.

Roads were almost nonexistent in the area, except for lumber roads built when the native pine and hemlock were cut a century earlier. There were few clearings, even on the mainland, and even these had been opened mainly to provide a vegetable garden or a hay crop to winter the family cow or horse owned by the impoverished year-round resident.

One could explore the lumber roads for miles inland without seeing a house or even a clearing. In our young minds we imagined that we were at the jumping-off place where to the north an endless wilderness extended to Hudson Bay and the arctic.

Young Aldo had such a dream; he longed to take a summer trip up a north-flowing river leading to James Bay or Hudson Bay. Looking forward to such a trip, he knew he must learn to handle a canoe, so he prevailed on Father to order one, which eventually arrived in a crate on the S.S. Islander. His was the first canoe in the area since the days of Indian dugouts or the birchbark canoes of fur trappers.

We all became proficient in travelling by canoe, not only in the small channels and bays near our cottage, but even along the shoreline of big Lake Huron. We were sometimes in such rough water that when our canoe was in the trough of a wave only the tops of trees on the shore could be seen. But we hardly ever shipped any water because we knew how to use our canoe. We sat on our heels on the bottom.

Hunting and Fishing

We took three- and four-day camping trips along Lake Huron's north shore, which led us to some good trout streams and some early-season hunts for partridge and a few black ducks. We never killed much game, but tried to live off the land. At times, we had to resort to stews of rice and potatoes, with an occasional red squirrel or flicker carcass to give a touch of meat. We found blueberries and sometimes beach plums to spice our pancakes with syrup.

Aldo, in his younger years, never realized his dream of canoeing a north-flowing river. This huge land he held as a blank spot in his mental map of the North Country. But that image only increased his interest. That mystery of a blank spot stirred his imagination and became the material his dreams were made of.

As a fisherman, Aldo was first a smallmouth bass man, then northern pike, followed by brook trout from the few little cedar swamp streams on the nearby mainland. Our top bass fishing area was known as Split Rock, named for a split boulder lying on the lake bottom in plain sight through six feet of clear water. The bass came and went through this area, feeding among the rocks.

The weed beds where Aldo caught his pike were near the bass grounds. Here, he learned to cast a Skinner spoon, which the pike struck. He never kept bass unless they weighed at least a pound and a half or pike if they were under three pounds. I recall him returning one evening with six big pike in tow on lines behind his skiff, rowing slowly so the pike could swim along without drowning.

His big fishing moment took place early one morning while trolling for the rare muskellunge; he landed one weighing 15 pounds by actual scale weight. A generation earlier muskies had been caught quite frequently here, but this was the last big one I know about.

Another quarry of Aldo's during these summers was the sora rail, a tiny bird that dislikes flying. The head of the big bay near our cottage grew shallow and the bottom muddy, a perfect place for wild rice if the water remained clear enough. The little rails made a living walking over the floating rice stems eating insects. I would row Aldo, who stood in the bow of the boat, through this rice bed. When a rail jumped, Aldo would try to shoot the bird. I don't recall how many he bagged, but they were few, despite the fact they would fly only a short distance before again alighting on the rice stems.

Varied Recreation

While Aldo spent most of his days in the woods or fishing, he by no means neglected the fairer sex. He enjoyed the evening dances and often took his partners on walks to enjoy the northern lights or the moonrise from the Mile Walk. After the canoe arrived, he even took evening paddles in the canoe, where parents felt a chaperone was unnecessary. Mother, who loved and admired Aldo as a favorite son — a fact which all of us acknowledged — often worried about his youthful courting of college girls. But these were summer romances, and each was soon forgotten.

Another summer sport was sailing. We did not own a sailboat, but dear friends had a sloop, called the *Hyac*, which they loaned to us over a series of years. We launched and maintained her and hauled her out in the autumn. Our group of four learned to sail quite well and dearly loved the sport. I recall Aldo bringing the *Hyac* home from Hessel, Michigan, two miles away, where it had dropped its rudder. He steered by manipulating the jib and mainsail to achieve his desired direction. It took a real sailor to do that. In all our younger years none of us owned a gasoline driven boat.

Among the most remembered camping trips of our youth was a trout fishing trip to Taylor's Creek, about eight miles inland from Hessel. To get there Father hired a horse-drawn sled that could negotiate the rugged logging roads in summer. In winter the loggers sledded their logs out on snowclad roads, which were often made slick by sprinkling the snow with water to form ice and make the heavy loads easier to pull. The roads were laid out to follow a downward slope to the water's edge, where in spring the logs were formed into rafts for towing or loaded onto sailing schooners, which carried them to the lumber mills.

Our eight-mile trip in took a long half day. Our tents were pitched above the little stream, which was small enough to wade easily and was often clogged with fallen logs. It contained a few deeper holes, up to two feet deep perhaps. I don't recall our having any protection from blackflies or mosquitoes, such as a mosquito bar, but we did have a concoction purchased from Von Lengerky and Antoine in Chicago called Lolly-ca-pop. This salve's effective period was short-lived. Flit and such repellants were far in the future.

We all caught some dandy brook trout — all we could eat, plus some to take home with us. By the second day, however, Father's eyes were swollen nearly shut and it was time to leave for home again. This trip taught us to hold off our camping dates until late August or early September when the scourge of blackflies and deer flies was over.

Another annual trout fishing trip was to Steele Creek, which emptied into the big lake about five miles north of our cottage. We rowed there in the early morning and returned in the dusk of the long summer twilight.

The stream was about the size of Taylor's Creek, but more accessible. It flowed through a meadow for its last half mile to the lakeshore after emerging from the usual dense forest of mostly spruce, balsam, and occasional aspen and balm of gilead.

On one such trip we were amazed to find that beaver had dammed the stream in the forested area and raised the water level nearly two feet, flooding a considerable area. We soon found ourselves forced to wade in the water. But the trout had responded to the improved habitat, and we made a record catch on that red-letter day, feeling well paid for the long row and wet feet.

As mid-September arrived, it was time once again for school. Often, we capped our season with a short canoe and camping trip along the wild north shore of Lake Huron. Then came the journey home and another year of school.

All of us appreciated our days at Les Cheneaux, not realizing that life's realities, particularly the need to enter the workaday world, would mean no more summer-long vacations.

Leopold family at the Shack
(front row, from left: Nina, Estella Jr., dog Flick,
Starker; back row, from left: Aldo, Estella, Luna)

11

Built on Honor to Endure: Evolution of a Leopold Family Philosophy

Sharon Kaufman

Early one autumn morning, about the turn of the century, a man and his son arise and dress by the dim light of a gas lamp. Believing in a stick-to-the-ribs meal, they breakfast on last night's pork and beans, then prepare a light lunch and head out the door with one gun and one stick in hand. They catch the early train from Burlington, Iowa, crossing the Mississippi River to Illinois where the bluffs are set back and the land is a broad expanse of wetlands. The ride costs them 12 cents each.

Arriving just before sunrise at the hunting club where the family has had a membership for a number of years, the party of two hikes toward the lakes and swamps where migrating ducks are sure to have spent the night. It is a typical duck hunting day—a cold chill is carried by a north wind under cloudy skies that threaten snow. Early hours are spent jumpshooting or flushing the ducks from their night's resting place, the father spending as much time carefully watching his son as watching for ducks.

A bird takes to the air and the boy jumps back startled, while the father folds it up with the single shot from his gun. The boy beams with pride as the father attaches it to his belt and continues on.

At lunch time they sit for a few minutes, sheltered by some pin oaks at the east end of the club grounds. A piece of bread and a hunk of dried meat are enough to carry them through the afternoon.

They spend much of their time just looking around, comparing the sightings of game and the advance of the seasons to previous years. They find a mink's den at the base of a tree and speculate as to what the mink might be hunting. All the time the father's watchful eye is on his son, and he's wondering: Is he ready for the next step in his training, the privilege of carrying a real, if unloaded, gun? Does he understand why they're hunting the way they are?

By evening they have five or six mallards and maybe a black duck or canvasback, plenty to provide a meal or two before the meat spoils in this era of no refrigeration.

They get off the train just after dark. The boy's arms are weary from carrying his stick all day, but he wants to carry the game the last 100 yards up the hill to the house. With pride, he sits down in front of the fireplace, smooths the feathers and lays each bird out to be admired, while the events of the day are recounted and recorded. The gun is carefully cleaned, oiled, and returned to its proper rack.

Plenty tired, physically and intellectually, they bathe and go to bed.

Built on Honor to Endure

The father's name was Carl Leopold, son of German immigrants and manager of a wood office furniture manufacturing firm in Burlington, Iowa. As was customary in the late 1880s, Carl Leopold chose an advertising motto for his company that represented his ideas about the company and the product. "Built on Honor to Endure" not only described his business ethic, but also his attitude toward his family, home, and the environment. Each of his sons and his daughter learned from their father the whys and hows of hunting and enjoying the outdoors. Carl Leopold's approach toward that education was not to teach but to share. By setting an example and asking why, he helped them develop a sense of wonder, a curiosity, and a desire to understand why. He possessed a gift of foresight and the patience required to guide his children in discovering for themselves the changes that were occurring around them.

The Leopold Family in Burlington

Carl Leopold was the second generation of his family to live in Burlington. The first generation was represented by Charles Starker, Carl Leopold's uncle. Starker, a professional architect, was sent to Burlington from Chicago in the 1850s to design and oversee the construction of a home for the gentleman who would become the third governor of Iowa. Starker was born and educated in Germany. He became so fond of the Burlington area that he remained there following the project, despite the fact there was limited work for an architect. Proud of his German heritage and work ethic, Starker amassed a sizeable estate as a merchant.

Charles Starker purchased his first home in Burlington, at 101 Clay Street, atop the Mississippi River and overlooking the Illinois wetlands. For years the only transportation to Illinois and the Crystal Lake Hunting Club, of which Starker was a founding member, was the train.

Starker extensively landscaped the exterior of his home with exotic trees, shrubs, fountains, and flowers. He employed a full-time gardener. The interior, with its tall ceilings, was decorated with gas lamps, candles, and mirrors. Stories of the Starkers' traditional German Christmas Eve celebrations are full of memories of ceiling-high Christmas trees decorated with glass ornaments and real candles. The light of the candles is said to have been reflected in a dozen mirrors.

Starker's architectural talents were delegated to such spare-time civic projects as the design of the opera house, Aspen Grove Cemetery, and the public library. One of his most controversial projects was Crapo Park, Burlington's largest park. Burlington at the time was a tiny island surrounded by what many considered untamed wilderness. Imagine Starker convincing landowners in a 10-year-old community of the need to donate land for a park. Crapo Park's central lake was named Lake Starker for its principal designer.

Charles Starker was married in Burlington. He and his wife had seven children, five of whom died during their second summer from a bowel upset known as "summer complaint." A son, Arthur, died as a young man, leaving one of the seven, a daughter, Clara.

Clara Starker married her first cousin, Carl Leopold. They moved into the family home and lived there until 1894, during which time Clara gave birth to three children, Aldo, Marie, and Carl. Thoughts of a fourth child prompted Clara to ask her adoring father to design and build a home for them in the cow pasture. A fourth child, Frederic, was born in 1895 in that home at 111 Clay Street and lives there today.

Aldo is, of course, the best known of the family. From a chicken coop on an abandoned farm in south central Wisconsin, he wrote *A Sand County Almanac,* in which he spoke of a new attitude toward the land—plants, animals, water, and soil. His land ethic represented an extension of his values, his sense of right and wrong, to include the entire land community.

Aldo's shack was his special wild place to which he could go to escape the day-to-day tasks of a university professor. It is maintained today in a near-original condition and provides housing and inspiration for graduate students doing research on the Leopold Reserve.

Aldo, his wife Estella, and their five children, worked hard to restore the area to its original condition. The wetlands and prairie provide a lasting memorial to a man from whom we've received so much.

Aldo Leopold died in 1948 fighting a grass fire on a neighbor's property. He was buried in the Burlington cemetery designed by his grandfather.

Aldo's younger brother Carl married the sister of Aldo's wife and succeeded their father as manager of the family business. Carl died in 1954 after an extended illness.

Duty, family, and a true enjoyment of the business world prompted Frederic to take the reigns of the family business upon his brother's death. Although he enjoyed his work and was very successful, Frederic loved the outdoors and possessed a profound understanding of the interrelationships within the natural world. Although less famous than his brother, Fred exemplifies the family values and is himself an accomplished wildlife researcher. His story bears telling.

Aldo Leopold recognized his brother's skills at observation and perception and encouraged Fred to become involved in wildlife management and research projects. One of Aldo's graduate students, Art Hawkins, approached Fred in the late 1930s and suggested they start a wood duck nesting box project at the Crystal Lake Hunting Club, of which Fred was president. The first year of the project was largely unsuccessful. But early one morning in the spring of the second year, Fred and Art watched as a pair of wood ducks in Fred's yard exhibited behavior typical of wood ducks searching for nesting sites. They immediately placed three boxes in the trees in the yard. All three boxes were used that year, marking the beginning of Fred's wood duck nesting project.

Fred's family, his late wife Edith, and his daughters Edith and

Margaret, shared Fred's love of the out-of-doors and his research. Margaret's son, Jim Spring, has been the focus of much of Fred's attention and today climbs the ladder collecting wood duck data.

Information gathered by Fred includes pioneering documentation of the behavior of the wood duck hen when choosing a nest site, which sometimes takes several days; building the nest itself from her own breast down; laying the eggs; her behavior and her relationship with the drake during incubation; and the pipping and hatching of the eggs. From a blind, Fred recorded on paper and film the behavior of the ducks on the day of exodus from the nest. Imagine a businessman beginning what would have already been a full day by watching the female leave the nest and call softly from the ground, 10 to 15 feet below. He observed the chicks, less than 48 hours old, jump from the nest and follow the hen down a 120-foot bluff to the river bank and across the three-quarter-mile-wide Mississippi River to the rearing grounds on the Illinois side. Fred often assisted the ducks by removing potential predators from the yard or by shepherding them across railroad tracks and the river. There were other nest users, such as screech owls, that were welcomed; these provided interesting information themselves.

Fred's numerous scientific journal articles about his research and his hundreds of presentations to youth and civic groups on the habits of the wood duck are well known. In 1966 he received the Wildlife Conservation Award from the Iowa Wildlife Federation, in cooperation with the National Wildlife Federation and the Sears Roebuck Foundation, for outstanding contributions to the wise use and management of the nation's natural resources. All of this has earned for Fred the honorary title of "Mr. Wood Duck."

Evolution of a Family Philosophy

Many members of the Leopold family are well known for their high degrees of knowledge and ethics toward the environment. This is the result, at least in part, of family influences. These family traditions have implications for environmental education and for parenting.

Through several generations of the Leopold family, the hunting experience has been the primary means of introducing new generations to the natural world. What distinguishes this hunting experience from hunting as traditionally defined is that to the Leopolds hunting is the pursuit of understanding of the natural world. The quest for game is but a subset of that experience. Success in hunting is measured in the amount of understanding and enjoyment gained. Returning with game is fine, but it is not a primary measure of success. Leopold hunting requires curiosity, questioning, and an intelligent appreciation of the natural world. In other words, the focus has been transferred from killing to the actual hunt itself.

The following excerpts from my "Leopold Family Oral History" (the Des Moines County Historical Society Museum, Burlington, Iowa) may serve to clarify this point.

Nina Leopold Bradley, speaking of her grandfather, Carl Leopold:

> I think he subtly indoctrinated all his kids with very strong principles of conservation. And it was very subtly done. It was not hitting you over the head, "You don't shoot ducks in the spring," but he would say, "I'm not shooting ducks in the spring because I think the population of ducks is going down, and this is a very bad time to slaughter them." By his own example, he formed children with a very good sense of aesthetics and conservation. [Tape no. 3]

Aldo's youngest daughter, Estella, writing of her father:

> You have probably heard from other members of the family the way in which Dad taught us the lore of hunting. Before being of an age when I was permitted to carry a gun, I used to follow Dad on hunts, trying to keep out of the way and following his directions on how to help the bird dog flush the game at the right moment.
>
> There was much stopping and discussing tracks and sign, droppings, trying to reconstruct what the animal had been doing, what it was eating, etc. It was both fun and a lot of work. At the time, I didn't feel too sorry for the birds — that came later. Whether we were hunting or not, long walks with Dad always involved these kinds of ecological analyses. I don't think he missed seeing much that was going on in the landscape, he was such an astute and knowledgeable observer. He knew every species of bird, plant, and mammal, and usually talked about them as individuals. All this made the biotic community very real and very interesting (exciting). Uncle Fritz is just like Dad in this respect.
>
> When I expressed an interest in carrying a gun, Dad gave me a wooden replica of a single barreled shot gun he had made, and for many trips I was told how to carry it, but not to point it at people ever, how to lift it to follow the flushed bird, etc. Then when he was

satisfied with my performance, he let me carry a real single-barreled shot gun (20 gauge, carrying only one shell) and his own double-barreled 20 gauge was always quick to follow my shot if he thought I had wounded the bird. I never graduated to a gun with two shells.

Through all of this, Dad talked much about what was okay to shoot at and in what season it was not too hard on the population (though I was never much of a threat). As you may have heard, the pinnacle of difficult sport hunting at the end of his career was hunting ruffed grouse in central Wisconsin. This was tough work, beating one's way through the brush in deep woods, fording creeks, and being scared stiff when a bird would rise like thunder and you have a few seconds to get the gun up, swaying to follow his path through the alder thicket. I never came close to one of these birds, but always loved tramping through the woods with Dad, and having picnic lunches back at the car with Mother and Dad. Eventually I ended up the way I began, helping the bird dog flush ruffed grouse and not carrying a gun at all. [Unpublished letter]

Finally, Jim Spring, speaking of his grandfather, Frederic Leopold:

I've been very fortunate to grow up with this character, because from day one, when we would go out walking, not just hunting, any time of the year, we were observing constantly. He always wants to ask, "Why!", which is a classic scientific approach. You don't just go out and take a walk, but you start looking at things and analyzing them and try to see the interaction that's taking place. You see a tree that grows up in the woods at a strange angle, and you can see, thanks to his observation, thanks to his help, that the tree didn't just do that by chance. There used to be a larger tree that's no longer there. It made the younger tree have to reach out from under it to get to the sunlight. And it may be a hundred and fifty years old now, but there used to be another tree there. And just little touches like that. I feel very lucky to have that kind of start relative to the out-of-doors.

It was always very subtle, never overt. For instance, handling a gun. When I first learned to handle a gun, I carried it empty for a long time and he was always watching me, subtly watching me. You took it very seriously!

It's really wonderful, I think. It's respect that is the underlying theme of his approach to life, really. It can take many different forms. He thinks everything through very thoroughly. [Tape no. 7]

There are many similarities in these stories: hunting; a one-to-one parent/child experience; observation and questioning not only about the quarry they sought but also its relationship to the rest of the natural world; strict training in the use of a deadly weapon; a true respect for each member of the biotic community; and, most of all, setting an example as an excited, curious sportsman.

I do not believe that these activities alone are responsible for the personal and professional accomplishments of the selected members of the Leopold family. Pursuing these ideas, however, may add to a theoretical foundation in the areas of environmental education and parenting.

12

The Leopold Memorial Reserve

Charles C. Bradley

The Leopold Memorial Reserve is a private wild area of almost 1,400 acres dedicated to the memory of Aldo Leopold. It contains his original 80 acres, now expanded by the family to about 267 acres. The rest is owned by cooperating landholders. A landmark of the original acreage is the Shack, now listed in the National Register of Historic Places.

The reserve is located in south central Wisconsin where the Wisconsin River and its floodplain cut a swath through the ground moraine with its bogs and marshes left by the last ice sheet. About two-thirds of the reserve is floodplain forest and marshland, dotted with ponds and laced with river sloughs. The remainder is hilly ground moraine covered by a mixed oak-hickory-pine forest, broken by a few fields still under cultivation. Underlying the reserve is a deep, sandy substrate that is infertile and highly erodible.

The area was settled by European immigrants in the 1840s. At that time oak savanna maintained by fire was a common ecotype. Lumbering, cultivation, drainage of wetlands, overgrazing, mowing, and fire suppression caused rapid changes in the vegetative cover and deterioration of the land. By the 1920s most farms had been abandoned to brush and weeds, wind and weather.

A Land Rehabilitation Project

In 1935 Aldo Leopold purchased one of these abandoned farms and started to reverse the process of land deterioration. His goal: to build it back to something like its presettlement condition when it supported a healthy crop of wildlife.

Reading *A Sand County Almanac* gives little indication of the research and planning, much less the sheer physical labor, that Leopold and his family put into their land rehabilitation project. By the time Aldo Leopold died in 1948, at least half of the 16,000 pines they had planted were flourishing in the sterile sand. The worn-out cornfield had become one of the first prairie restoration projects. By this time the Leopold children had grown and dispersed. The Shack had few visitations. The Leopold land was left mainly to its own considerable momentum. Wild things were finding new niches and moving in.

A Fitting Memorial

In 1967 Reed Coleman of Madison, the son of Aldo Leopold's good friend, Tom Coleman, created what is now called The Sand County Foundation. He was also managing a tract of land purchased by his father just south of Leopold's acres. With the growing threat of recreational development in the area, Reed made a counter move. He persuaded the other owners of land surrounding the Leopold tract to pool their land in a cooperative wild area memorializing Aldo Leopold. By 1968 the reserve was officially in place.

Among the landowners it was understood that their land would be under common management funded by The Sand County Foundation. A generalized management plan was drafted by one of Leopold's students, Robert Ellarson, who became ecological advisor to the foundation. Frank Terbilcox, one of the landowners, gave up his florist shop in Baraboo to accept the job of manager.

The goals of this new reserve were to continue Leopold's land rehabilitation program, to create diversified habitats for native Wisconsin species, to develop appropriate management techniques, and to initiate educational programs suited to the area and its management.

The reserve was not to be open to the public, but selected educational group visits, mainly for adults, could be arranged through the manager. To give stability and continuity to the area and its program, the foundation, in addition to funding the tract's management, retained first right of refusal when any of the tracts came up for sale. Otherwise, the landowners retained their normal rights and enjoyed, in addition, access to the reserve for recreation, hunting, and firewood supply. Because the landowners are officially part of the reserve management board, they are represented in all internal affairs and may participate actively in the management program itself.

In the few years after the reserve was created a network of trails was cleared, many ponds were dug, and prairie burning became a routine management tool. Many a wildlife manager and student profited from the group visits, to see what rehabilitation efforts were going on and to see the Shack, now famous through *A Sand County Almanac*.

Extending the Memorial

In 1976, eight years after start-up, Charles and Nina Bradley arrived at the reserve with the proposition that the memorial to Aldo Leopold was incomplete without graduate students and a research program. The Bradleys became co-directors of research and built the Study Center, with a laboratory and work space for students in the lower level.

The first four Leopold Fellows took to the field in the summer of 1978. Their assignment was to create a working base map and begin a comprehensive inventory of species and environmental factors. Later would come the studies of correlation and interrelationships that would ultimately help illuminate the ecology of the reserve and interlock with its management.

Leopold Fellows are selected annually by a committee on the basis of proposals submitted for review. Projects may be ones suggested by the committee, or one of a student's own choosing. If the latter, the project must at least have some relevance to the reserve itself.

Over the last eight years the reserve has been the focus of 27 fellows who have completed 42 reports, reprints, and theses based on their fieldwork. Four of these fellows have received masters degrees, and five have completed their doctorates.

In 1986-1987, as part of the Leopold centennial celebration, the reserve has added to the team a senior research fellow, Dr. William Karasov of the University of Wisconsin-Madison Department of Wildlife Ecology. Dr. Karasov's research into animal energetics represents the most complex and fundamental ecological field study yet attempted on the reserve. Also joining the centennial group as senior fellow-at-large is Dr. Grant Cottam, recently retired from the University of Wisconsin-Madison Botany Department and a former student of Aldo Leopold.

The Future

In the longer perspective of the land rehabilitation assignment, there are seven prairie or oak savanna restorations in full swing at the reserve. Forest and agricultural land rehabilitation are both underway as well. Educational traffic on the reserve — visitations, seminars, Fellows, and assistants — is approaching 1,500 person-days per year.

While numbers do not express quality, they at least suggest workload. Clearly, too, the reserve is growing more beautiful and healthy each year. (Recall that health and beauty are two of Leopold's requirements for an expression of the land ethic.) Finally, surrounded as it is by land devoted to other things, the reserve retains, as Leopold wished, a place for "things wild and free." Thanks to generous private grants and donations and to skillful management of funds by The Sand County Foundation, the reserve is in solid financial shape, which has permitted modest additions to the original acreage and an expansion of programs.

The reserve is probably unique as an example of private initiative in the reclamation and preservation of wild areas. This would have pleased Aldo, who saw the private landowner as the key to land health.

Has the Leopold Memorial Reserve helped the nation or, for that matter, any individual to accept the land ethic as a way of life and a key to the survival of genus *Homo*? The absence of any hard data on this matter permits our continued hope.

13

Reflections and Recollections

Carl Leopold, Estella Leopold, Luna Leopold,
Nina Leopold Bradley, and Frederic Leopold

Leopold the Person

Dad was as kind, considerate, and tolerant a person as any I have
ever known. Practically never did he criticize anyone personally,
even when he disagreed with that person. He treated even the most
humble with the same respect as the most distinguished. This was especially noticeable when travelling—the porter in the train, the
shoeblack, the waitress in a cafe—all were engaged in conversation,
in which he might ask about the person's interests, avocation, or
work.

I have the impression that the idea of an ethical view of land was a
gradual outgrowth of his concern for individual people, an extension of his innate feeling that all persons have good and interesting
qualities that must be understood and respected. Examples of this
interest in others is illustrated in his essay "A Man's Leisure Time."

Luna Leopold

I always marveled at the way Dad could talk to anyone and speak
in such a way that made them feel comfortable communicating with
him. He was always pleasant and direct, tactful and interesting.
When he was speaking with professionals in his field, he used elegant language; a certain number of technical words, though he usu-

165

ally avoided scientific "jargon"; and generally came right to the point. He asked a lot of questions and was a good listener.

With his friends socially he was much the same, but always added an element of good humor. He was great at telling stories that were to the point, or funny. He laughed a lot and was a person of good cheer. He was very interested in politics and what his friends thought of news developments.

With farmers and people in the country, he would often start the conversation by making observations about the weather, then ask the farmer how his farm animals and crops were doing, offer some observations about a pheasant or snipe he had seen while walking to their place, or ask how the native quail were doing this year. Finally, he would get around to asking about the matter of business that brought about his visit. The main business never came first! Furthermore, his speech was embellished with colloquial country usages reserved for relaxed family discussions or talks with farmers. This last feature made him sound a lot like the farmers he addressed. He was always interesting and entertaining; I cannot remember him ever being unpleasant.

Estella Leopold

Dad tended to avoid social events, especially in his later years. I suspect that being prone to waking up so early in the morning had something to do with that. But he also felt that social conversations at parties were usually boring. Mother, on the other hand, was very sociable and loved to go to parties. I can remember times when Mother would work really hard to persuade Dad that he should go to a party and then, when he had a good time, remind him of it the next day.

Carl Leopold

In her book, *Thinking Like a Mountain,* Susan Flader discusses Dad's reaction to his critics in the deer debate. My personal recollection is that Dad was deeply and personally concerned. His frustration at the public response to the deer problem was expressed at home in a kind of quietness. He would sit for long periods in his easy chair, thinking, his whole being consumed with the problem.

At the time, Dad was working with Paul Errington to publish their 20 years' study of the bobwhite quail. The two authors were unable to agree on wording, conclusions, analysis, etc. In a final attempt to reach agreement, Dad took the train to Iowa where he spent part of a week working with Paul. He returned with the decision that Dad would remove his name from the manuscript. Dad appeared disappointed and sad, but showed no rancor.

Always Dad tried to be persuasive, never arrogant. He had a very basic respect for other people's thoughts and never lashed out. He appeared calm and worked with persuasive logic.

Nina Leopold Bradley

People sometimes ask me whether my father was a religious man, and I say no, he wasn't, but he was a very moral person. To illustrate that I have a story, a recollection from long ago.

My father and I were working in the woods one time, at the Shack. We heard some sounds from across the sandhill. It sounded clearly like somebody digging; you could hear the clack of the shovel. We looked at each other and decided we'd better go see what was up. Following the sound, we came upon two middle-aged couples. The men were in baseball caps and sports shirts, the women were in bright, gaudy pant suits. They were digging up our pine trees. Our blood pressure immediately began to rise.

Dad confronted them, "What are you doing?"

One of the men responded, "Well, we're just digging up these pine trees."

Father countered, "Who said you could dig up these pine trees?"

"We're not going to kill them; we're going to take them home and put them in the yard; they'll grow there," explained one of the men.

At that, Dad tilted his cap back on his head and exclaimed, "Well, god damn!"

One of the men interjected, "Please, there are ladies present."

I really was expecting Dad to get so angry that he'd really blow, but at that moment he leaned back and just began to laugh the most marvelous, silly laugh.

Carl Leopold

It's great fun to look through his books today to find extensive penciled notes in the books that he took most seriously. He was a great reader of Thoreau; he was also very much impressed by Donald Culross Peattie. I'm not quite sure why because Peattie seems to me a little trivial. But Dad found great joy in his works, and I remember he and Mother would read them aloud. On the philosophical side, he was much influenced by the Russian, Ouspensky. His copy of *Tertium Organum* is still somewhere in the family with lots of notes.

Carl Leopold

Mother and Dad did a lot of reading out loud in the evenings. Dad was particularly enamored by western exploration accounts, such as Fremont's journals. In a lighter vein, when A. B. Guthrie's *The Big Sky* was printed, I can remember that was one of his favorites. It was like the journals of early mountain men, exploring the upper Missouri, describing rather beautifully the wildlife biota of the early western frontier.

Estella Leopold

Mother read to Dad almost every evening because Dad had problems with his eyes. As they got older, very often Mother would go to sleep while reading aloud, and Dad would sit there with this wonderful smile on his face, and finally say, "Well, let's go to bed!"

Nina Leopold Bradley

We heard a lot about Dad's father and his brothers, Frederic and Carl. I think they were all a very close family, and I think there was a lot of mutual intellectual inspiration and exchange among them, specifically in the outdoor realm.

Carl Leopold

He was a marvelous craftsman in terms of making his own archery equipment. There were several decades in his life when he was very active in archery; he made all his own bows, all his own arrows.

I can remember one time he was sitting in the living room putting the feathers on a set of arrows he was making. This is exacting work; you have to sit still, glue them just right, and clamp them onto the shaft of the arrow and then not move until they're set.

Nina and I were rough-housing. Dad gave us a couple of exclamations, "Would you kids please be quiet!" He was working away, and we continued to do whatever we were doing.

Finally, I remember Dad lifting the board that was on his lap, with all of his equipment, and putting it down on the floor, and I thought, "Oh my God." He got up and took my hand and Nina's hand, then walked to the coat closet, opened the door, and locked us in. To that moment, Nina and I had been enemies of the worst order; now we were companions in distress. We threw our arms around each other and wept bitterly.

I could hear Mother outside saying to him, "Aldo, I think you should let them out now."

My memory is that he forgot us, and we stayed in that closet for a very long time!

Carl Leopold

Especially in later years, Dad became sleepless early in the morning. He would wake up at 3:00 or 4:00 or 5:00 in the morning and finally get up in desperation. Being unable to sleep, he acquired the habit of listening to the birds as they began to chime in in the morning and that, of course, led to his famous bird phenology paper.

Nina Leopold

This is when he did his most creative writing, when he got to the office in the morning, before the secretary got there. He did go to bed very early, usually around 9:00, and it really didn't make a lot of difference what the company might be. Very often we would entertain students for dinner, and at 9:00 Dad would excuse himself and say he had to go to bed. Mother would be left with the students—

wild souls, some of whom really didn't know very well how to say good-bye or good night. Mother would sit up with them for hours!

Estella Leopold

Life at the Shack

Our introduction to Dad's farm came on a winter day in 1936 when he drove his family—his wife and five children and the dog—to have a look at the recently acquired farm. We drove the 50 miles in a February blizzard, shoveling out of snowdrifts and finally trudging the last quarter mile on foot through the snow. We arrived to find the only remaining building, an old chicken coop, complete with chicken perches and waist-deep with frozen manure. As far as the eye could see was corn stubble, cockleburs, broken fences, and blowing sand and snow.

Little enthusiasm was expressed by the family until, on subsequent weekends, we found ourselves involved in fixing up the old chicken coop for weekend living quarters. Materials for repair came not from the lumber yard but from the floodplain of the Wisconsin River, which bounded the north edge of the farm. I think he got this idea from his reading of Thoreau. Such treasures as old bridge pilings, 2x4's, planks of all widths and lengths became crucial material for benches, siding, tabletops, and window frames.

Windows and a door for the Shack my father found in the local dump. Today, I find myself smiling as I contemplate Dad, the sophisticated university professor, scrounging through the dump, looking for usable objects!

Building a fireplace involved the entire family as we hoisted a huge limestone block into place as a lintel. Dad cut a red cedar from the moraine above the Shack, hand-hewed boards from the log, and with a lindseed oil finish made a handsome mantle. As carefully engineered as was this fireplace, it smoked in the 1930s and it smokes today.

Dad's selection of sick land as a place for his family outings was perhaps a new concept in recreation. The land our father purchased for $8.00 per acre had been abused, misused, destroyed. It had been carelessly lumbered, carelessly farmed, and carelessly abandoned. It had nothing. Even after the spring thaw, the broken, sagging fences repeated the story of misuse and despair. This kind of farm was not hard to find in the 1930s at the end of many years of drought.

And here was the challenge. Would it be possible to bring it back to health?

The marathon began. Over a period of 12 years, we slaved with our father and mother. We gathered seeds of native prairie grasses and flowers from nearby areas; we planted them among the old corn stubble. We planted native pine trees—whites, reds, and jacks. We planted native hardwoods and forest wildflowers and shrubs. We planted, then carried pails of water. We planted some more. We learned how to nurture, how to care.

Mixed with these weekends of hard work was plenty of fun and excitement—censusing woodcock at dusk, singing before the fire at dark, watching the oak coals at night as they warmed our shins. Family camaraderie grew and expanded—never to die.

Nina Leopold Bradley

The Shack was a prime place for interaction with his family. All family members participated in projects.

It is mysterious to me how he transfused each of the family members with such enthusiasm for the projects at the Shack. I can remember my girl friend being indignant and hurt that I so often wanted to go to the Shack instead of staying in Madison on weekends. I never felt that I was under any pressure to go to the Shack; I simply wanted to be there.

Carl Leopold

Mother was a musician; she had played the piano as a youngster, and she sang. She taught us children simple songs in Spanish. Some of us learned broken Spanish, her native tongue, as a result.

Dad was musical, I believe, but he was so shy that he almost never sang. I only heard him carry a tune once. That was when he whistled a tune to us to tell us that he and Starker had just come back from a turkey hunt in the Southwest. This Mexican tune we had to learn, but he wasn't going to sing it; he was going to whistle it. Starker provided the words.

Mother was the music inspiration in the family, though not necessarily the only talent. Carl, my brother, and others of us learned to play the guitar. We ended up singing around the fireplace at the Shack, Spanish folk songs and other kinds of ditties from folklore; it was lots of fun.

Estella Leopold

One of the things Dad had picked up from Mother was a family expression of "whoopsie." This became then "whoopsie-daisy" when anything got dropped.

One day we were up on a long ladder repairing the top of the Shack's chimney. I was holding the mortarboard. Dad, while putting the bricks in, dropped a slab of mortar. When I said, "Whoopsie," Dad countered, "Whoopsie-daisy, god damn it!"

Carl Leopold

The Leopold Influence

The combination Mother and Dad used of setting the good example and being really interested in their children was effective. In most important things, such as picking a career area, they did not try to lead us, but listened quietly to our ideas and made gentle suggestions. For example, I had noted that my elder siblings had chosen study fields of wildlife, geology, botany, and geography, respectively, and I was bound and determined not to follow any of their footsteps; I wanted to do something different. Consequently, when I was in high school, I announced to Dad one day that I wanted to become an entomologist. He listened quietly, looking tolerant, then said, "I'll tell you something; on Monday I'll go down to the book store and buy you a copy of *The Spring Flora of Wisconsin* and a hand lens, and let's see what you can do with them." So he did. The following weekend at the Shack I went off armed with these new tools and became hooked on identifying plants. Somehow that did it, and I've been into botany and ecology ever since. There's a lot more to the story, of course — I would carry the plants I had identified back to Dad, and we would discuss whether I had gotten them right. This kind of reinforcement helped spur me on.

Estella Leopold

He avoided any clear efforts to steer us professionally. I can remember a conversation with him when we were driving somewhere in the car, asking him what he thought might be a good major for me at the university. He never did say what he thought, but when I mentioned botany, he was clearly enthusiastic about it. That clinched it for me, and I majored in botany.

Carl Leopold

Aldo had a great influence on me, especially as I grew older. Not that he had a lesser influence when I was young, but I didn't realize the influence he was having on me then. As I grew older and a little more sophisticated, I realized what an opportunity it was for me to be associated intimately in the family with a man of his talents.

Frederic Leopold

A Legacy

The advent of his serious essays aroused a new respect for him, but the thought of his leading a new philosophical movement did not really enter my mind until some time after his death. He has had a major impact on the way people think, with his land ethic in *A Sand County Almanac*.

Carl Leopold

Concern for my father's influence on the general public was farthest from my mind as Dad sent us early drafts (on yellow paper) of the individual essays for *A Sand County Almanac*. My personal reaction at that time was that the essays must have been written personally for us — for Mother and for Dad's children! The thought of outside response did not actually occur to me until the volume reached the book shelves after Dad's death. For 20 years after publication of the book (1948-1968), Dad's audience grew. During the second 20 years (1968-present), his audience expanded exponentially. Now, as the book is apparently being read by philosophers and "men of the cloth," Dad has reached a new and desired audience.

Nina Leopold Bradley

If He Were Alive Today...

He would be revolted by the actions of the Reagan Administration. Exploitation is the dominant position of that administration, and any concept of stewardship of our resources is held in contempt, real contempt.

And, he would be infuriated and sleepless over the madness of the arms race. He was a perceptive, careful student of political events; he had watched what militarism had done in Germany, even among sensitive and intelligent people. To see our own country submerging the talents of science into a militarization of the academic world would have been an intolerable burden. The madness of the arms race and the pretexts by the American president of wishing for a limitation of the arms race would have angered him for its hypocrisy and would have profoundly discouraged him as a threat to civilization. The present political situation of military confrontation would have been almost more than he could bear.

Carl Leopold

His health and outlook became depressed at the end of the war when the atomic bomb was used. That pessimism I think would have increased as he saw the effects of clearing of tropical forests, the growth of agribusiness, acid rain, and toxic pollution.

Yet he was a realist. His manner of teaching reflected the fact that various people have different ideas and he was always seeking to know better the viewpoint of the other person. To his children, and to others, his approach was to show by example — by the ethical way he lived his own life.

Luna Leopold

I don't think Aldo would ever be hopeless. I think he would always be optimistic that there is potential for education and that education can eventually produce results of wisdom, better guidance, better planning, better understanding. Aldo was never a pessimist; he was an optimist. He thought that people basically had good traits, that they could be educated; and he did his damnedest to do his part to broaden the viewpoint of Mr. Ordinary Man through his writings, through his understanding, and through his messages to the public on ecological subjects. I think he would still feel it was worth the battle of trying to educate the public for an appreciation of those values which are long-term, important values to our civilization.

Frederic Leopold

I take some comfort in the thought that my father is spared some of the momentous problems of the 1980s. The extravagant use of resources for an ever-expanding military force and their use to subsidize the high level of materialism of our culture, damming our western waters to support expanding conurbations, the gamble of arming the world with nuclear bombs are but a few items I am glad my father does not have to consider. His guiding words, written before 1948, are as relevant today as they were when he wrote them: "Examine each question in terms of what is esthetically and ethically right.... A thing is right when it tends to preserve the integrity, stability and beauty of the biotic community. It is wrong when it tends otherwise."

Nina Leopold Bradley

Foreword, 1987

The story of the ever-widening ripple of Aldo Leopold's influence on American thought constitutes one of the inspiring episodes of 20th century conservation history. When Leopold died four decades ago, he was among the seminal thinkers of his day. His greatness was only appreciated by a modest circle of admirers, however, because his most significant work—the incisive essays known to us as *A Sand County Almanac*—had not entered our literature.

In 1973, exactly a quarter of a century after Leopold's death, the editors of *Not Man Apart*, published by Friends of the Earth, asked various authors, scientists, politicians, philosophers, and environmental activists to list the five books they thought should be enshrined in an "Environmental Books Hall of Fame." The biggest vote-getters by far were *A Sand County Almanac* and Rachel Carson's *Silent Spring*. Historian Wallace Stegner later validated this assessment by predicting that when the record of "... the physical and spiritual pilgrimage of the American people" is compiled, Leopold's book will be "one of the prophetic books, the utterance of an American Isaiah."

What added an extra dimension to Leopold's contribution is that, although he built on philosophical foundations laid by Thoreau and Muir, he thought and wrote as a trained scientist who studied the impacts of trends that were shattering the age-old relationships between earth's inhabitants and their environment. His insights have a special resonance for us because he was a man of our century who participated in the action of his times.

177

We esteem this man as a pioneer ecologist, as a thinker who turned an abstract idea into a concept of wilderness preservation, and, above all, as a gifted writer who used spare, eloquent prose to formulate an imperishable land ethic for humankind.

All of us who are part of the environmental community are indebted to the perceptive folk at Iowa State University who not only arranged an Aldo Leopold centennial celebration but enlarged the Leopold canon by collecting and preserving this volume of essays.

Stewart L. Udall

Foreword, 1995

Midway between John Muir and Rachel Carson, during the depths of the Dust Bowl, an important chapter in the American conservation movement was written. For the first time, our nation recognized the importance of conservation on private land and acted on it.

No one spoke more to and about the connections among people and their land than Aldo Leopold. His thinking, his writing, and his teaching, along with the missionary zeal and action of the first chief of the new Soil Conservation Service, Hugh Hammond Bennett, inspired a stewardship ethic that has put its roots deep into the American landscape.

In a 1939 essay on the farmer as a conservationist, Leopold wrote "It is the American farmer who must weave the greater part of the rug on which America stands." Thanks to Leopold, Bennett, and the tens of thousands of individual landowners they worked with, the American landscape is healthier today than it was then. And many who have become part of that effort over the years are today weaving into that carpet what Leopold called "the colors which warm the eye and heart" as well as the "sober yarns which warm the feet."

As the essays in this wonderful collection point out, Aldo Leopold drew on John Muir's reverence for things wild and free, and he laid the foundation for Rachel Carson's plea for understanding of the interconnectedness of all life and prudence in our interactions with it. Leopold wrote like no other person about these themes: about wildness and wild life, about the need for caution when "tinkering" with natural systems, and about the importance of honest science to help us understand land health.

Most of all, Leopold challenged us to understand and influence human behavior. "All the acts of government," he wrote, "are of slight importance to conservation except as they affect the acts and thoughts of citizens." At the core of the land ethic is his challenge to understand wise land use not as a limit to freedom, but rather as "a positive exercise of skill and insight" born out of a warm and personal understanding of our relationship to the land.

This is no small challenge. In fact, it seems impossible. Yet we have no choice but to try.

This collection of essays about Leopold and his legacy contributes to our understanding of the task ahead. I value it. There are few authors I choose to read over and over again. Nor are there many authors whose ideas merit continuous thought and discussion. Aldo Leopold is an exception.

Thank you, Thomas Tanner, Iowa State University, the Soil and Water Conservation Society, and the host of fine contributors for giving us this treasure. Leopold would, no doubt, be honored by this book. His true legacy, however, should not be measured by words, but rather by actions inspired by these words. His legacy is written every day across our land. May this collection inspire us all to carry on.

Paul W. Johnson